R. INGRAM

*Advances in*
*Underwater Technology,*
*Ocean Science and*
*Offshore Engineering*

*Volume 32*

# *Subsea Control and Data Acquisition*

*For Oil and Gas Production Systems*

## ADVANCES IN UNDERWATER TECHNOLOGY, OCEAN SCIENCE AND OFFSHORE ENGINEERING

*CONFERENCE PLANNING COMMITTEE*

J. Cattanach, *Mentor Engineering Consultants Ltd.* (Chairman); L. Adriaansen, *Veritec a.s.*; C. Curran, *JP Kenny & Partners Ltd.*; P. Duncan, *PD Controls;* R. Morton, *SUT*; J. Pritchard, *SUT*; D. Scott, *Shell U.K. Exploration & Production.*

*Advances in*
*Underwater Technology,*
*Ocean Science and*
*Offshore Engineering*

*Volume 32*

# *Subsea Control*
# *and Data Acquisition*
## *For Oil and Gas Production Systems*

Papers presented at a conference
organized by the Society for Underwater Technology
and held in London, UK, April 20–21, 1994

KLUWER ACADEMIC PUBLISHERS
DORDRECHT / BOSTON / LONDON

Library of Congress Cataloging-in-Publication Data

```
Subsea control and data acquisition for oil and gas production /
  edited by Society for Underwater Technology.
      p.   cm. -- (Advances in underwater technology, ocean science,
  and offshore engineering ; v. 32)
    Co-sponsored by the Institute of Measurement and Control.
    ISBN 0-7923-2779-9 (acid-free paper)
    1. Oil well drilling, Submarine--Congresses.  2. Drilling
  platforms--Congresses.  3. Oil wells--Management--Congresses.
  4. Gas wells--Management--Congresses.   I. Society for Underwater
  Technology.  II. Institute of Measurement and Control.  III. Series.
  TN871.3.S894   1994
  622'.33819--dc20                                          94-7723
```

ISBN 0-7923-2779-9

Published by Kluwer Academic Publishers,
P.O. Box 17, 3300 AA Dordrecht, The Netherlands.

Kluwer Academic Publishers incorporates
the publishing programmes of
D. Reidel, Martinus Nijhoff, Dr W. Junk and MTP Press.

Sold and distributed in the U.S.A. and Canada
by Kluwer Academic Publishers,
101 Philip Drive, Norwell, MA 02061, U.S.A.

In all other countries, sold and distributed
by Kluwer Academic Publishers Group,
P.O. Box 322, 3300 AH Dordrecht, The Netherlands.

*Printed on acid-free paper*

Printed in the Netherlands

# Society for Underwater Technology

The Society was founded in 1966 to promote the further understanding of the under-water environment. It is a multi-disciplinary body with a worldwide membership of scientists and engineers who are active or have a common interest in underwater technology, ocean science and offshore engineering.

## Committees

The Society has a number of Committees to study such topics as:

Diving and Submersibles
Offshore Site Investigation and Geotechnics
Environmental Forces and Physical Oceanography
Ocean Resources
Subsea Engineering and Operations
Education and Training

## Conference and Seminars

An extensive programme is organized to cater for the diverse interests and needs of the membership. An annual programme usually comprises four conferences and a much greater number of one-day seminars plus evening meetings and an occasional visit to a place of technical interest. The Society has organized over 100 seminars in London, Aberdeen and other appropriate centres during the past decade. Attendance at these events is available at significantly reduced levels of registration fees for Members or staff of Corporate Members.

## Publications

Proceedings of the more recent conferences have been published in this series of *Advances in Underwater Technology, Ocean Science and Offshore Engineering*. These and other publications produced separately by the Society are available through the Society to members at a reduced cost. A careers pack 'Oceans of Opportunity' has been produced by the Society in response to the growing demand by students schools and colleges for up-to-date information.

## Journal

The Society's quarterly journal *Underwater Technology* caters for the whole spectrum of the inter-disciplinary interests and professional involvement of its readership. It includes papers from authoritative international sources on such subjects as:

Diving Technology and Physiology
Civil Engineering
Submersible Design and Operation
Geology and Geophysics

Subsea Systems
Naval Architecture
Marine Biology and Pollution
Oceanography
Petroleum Exploration and Production
Environmental Data

An Editorial Board has responsibility for ensuring that a high standard of quality and presentation of papers reflects a coherent and balanced coverage of the Society's diverse subject interests; through the Editorial Board, a procedure for assessment of papers is conducted.

**Endowment fund**

A separate fund has been established to provide tangible incentives to students to acquire knowledge and skills in underwater technology or related aspects of ocean science and offshore engineering. Postgraduate students have been sponsored to study to MSc level and subject to the growth of the fund it is hoped to extend this activity.

**Awards**

An annual President's Award is presented for a major achievement in underwater technology. In addition there is a series of sponsored annual awards by some Corporate Members for the best contribution to diving operations and oceanography, and for the best technical paper in the Journal

**FURTHER INFORMATION**

If you would like to receive further details, please contact
Society for Underwater Technology, The Memorial Building, 76 Mark Lane,
London EC3R 7JN.
Telephone: 071-481 0750; Telex: 886481 I Mar E G; Fax: 071-481 4001.

# *Contents*

**Session 1
Experience**

# THE REMOTE CONTROL SYSTEM FOR THE LILLE-FRIGG HIGH PRESSURE, HIGH TEMPERATURE SUBSEA DEVELOPMENT

E. GRUDE
Subsea Systems Dept.
Elf Petroleum Norge A/S
Stavanger
Norway

## ABSTRACT

This paper describes the main features of the Lille-Frigg subsea development, with particular emphasis on the subsea control system. Operated by Elf Petroleum Norge A/S, Lille-Frigg is a subsea satellite extension of the main Frigg Field. The reservoir contains gas and condensate at a high temperature and pressure, which requires the use of special high pressure downhole equipment and remote control components.

The three high pressure wellheads are located in 115 m water depth and feature a wellhead shut-in pressure of 530 bar and a wellhead flowing temperature of 110°C. The wells are tied back to the Frigg TCP2 platform about 20 km away.

These demanding operating conditions resulted in the development of a dedicated subsea control system, implementing some advanced technologies such as very high pressure hydraulic components (760 barg) and specific ROV tooling.

The subsea system is designed for initial installation using a combination of diver and diverless techniques. Diverless intervention is feasible in connection with change out of sub-systems such as Xmas trees, choke inserts and well control pods.

Extensive qualification programmes were initiated prior to the development phase in order to verify and also improve the available technology to meet with the stringent requirements of the Lille-Frigg needs.

*Volume 32: Subsea Control and Data Acquisition, 3–25.*

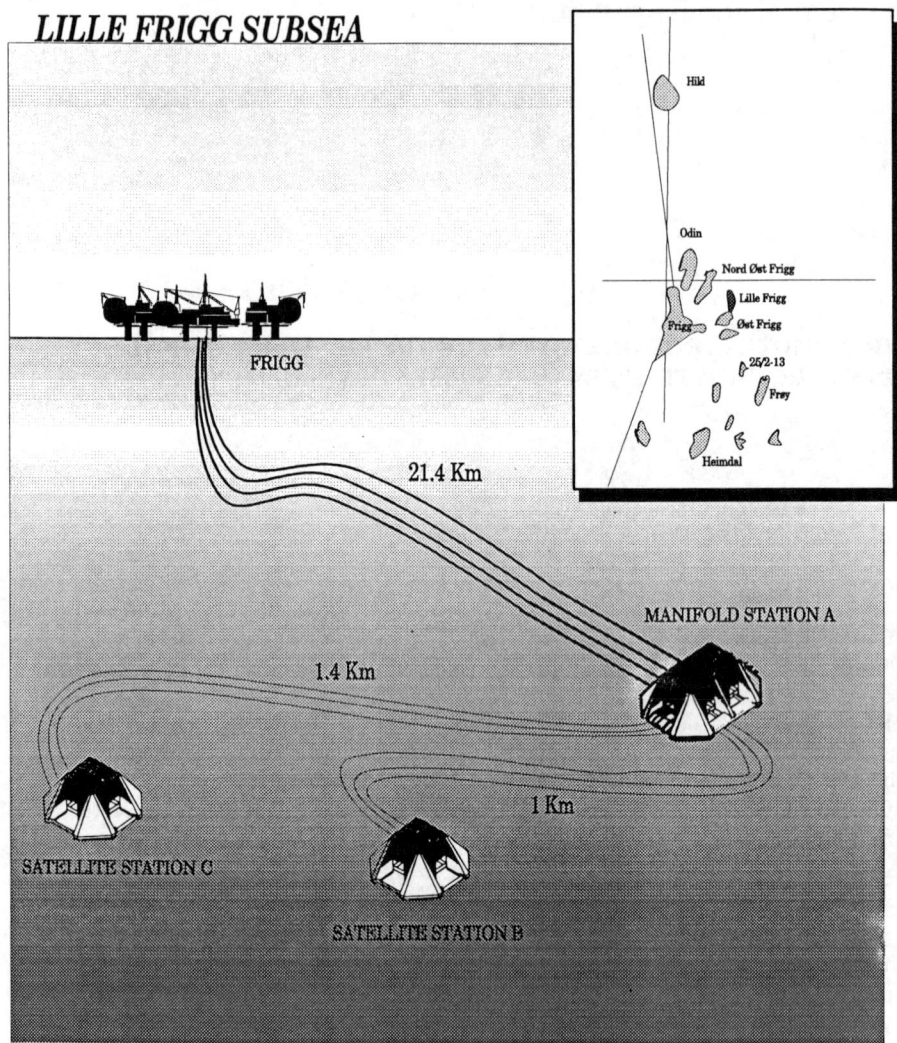

FRIGG AREA DEVELOPMENT
Fig. 1      LILLE FRIGG SUBSEA

## LILLE-FRIGG FIELD DEVELOPMENT

The Lille-Frigg field was discovered by the exploration well 25/2-4 drilled in 1975 in the northern part and confirmed by the appraisal well 25/2-12 & 12A drilled in 1988 in the central part. The field is located on the Norwegian shelf about 20 km north-east of the Frigg Central complex.

The reservoir pressure and temperature are 670 bar and 125°C at the gas/water contact level (3678m/MSL).

Recoverable reserves are estimated to be in the order of $8,5 \times 10^9$ std m$^3$ of wet gas and $3,5 \times 10^6$ m$^3$ of condensate.

The concept selection studies for Lille-Frigg included three basic principles for development:

- a wellhead platform concept
- a hybrid concept (similar to North-East Frigg)
- a subsea concept

The selected development scheme (see Fig. 1) is based on 3 vertical subsea wells producing through one 10" flowline to Frigg treatment and export facilities located on the TCP2 platform. Dry gas is exported through the Frigg to St. Fergus pipeline, while stabilized condensate is exported through the new Frostpipe line (Frigg to Oseberg pipeline). Monitoring and control of the wells is performed from the Frigg Field through an electro-hydraulic remote control system.

A central subsea manifold is located in the same structure as one of the wellheads "A". The two other wellheads "B" & "C" are located in satellite protective structures and tied in to the central manifold by flexible production lines, service/inhibitor injection and remote control lines.

Provisions have been made in the central manifold and also in the electrical and hydraulic distribution networks for extension of the subsea system with one or two additional wells if later required by production constraints.

## PROTECTIVE STRUCTURES

The three wellheads and manifold are protected by three separate structures:

- Protective Structure "A" - Manifold Station "A"
- Protective Structure "B" - Satellite Station "B"
- Protective Structure "C" - Satellite Station "C"

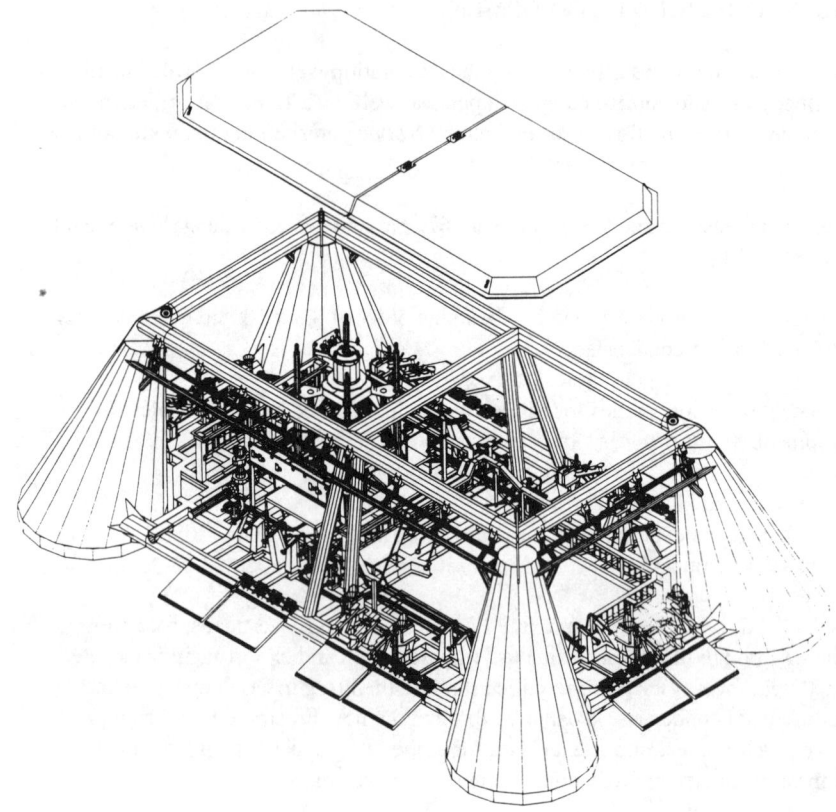

Fig. 2      LILLE FRIGG SUBSEA
            MANIFOLD STATION "A"

All protective structures are A-shaped steel structures with four circular corner structures. Upper and lower levels of the corner structures are connected with horizontal tubulars. The corner structures is filled with concrete to ensure adequate stability by gravity. The top of the structures is covered with hinged aluminum roof hatches to provide free vertical access to Wellbays and Manifold Bay when in open position.

Upper part of inclined vertical openings of the structures is equipped with hinged Trawl Board Deflectors. These are to take the impact from a hitting Trawl Board and tilt over to release the Trawl Board from the structure.

The wellhead is independently located in the Wellbays (7 x 7 meters free opening) surrounded by the deck structure.

The Manifold Station "A" is extended to also include the Manifold, the Electrical Dispatcher Unit (EDU) and two Hydraulic Distribution Units (HDUs).

The structures are equipped with the required inboard hard piping, isolation valves, Wellbay flexible jumpers, electrical cables, hydraulic hoses and tie-in/connection equipment for operation of the field.

All tie-in/connection bays are equipped with Sealine Ramps from structure to seabed for supporting of flexible sealines and umbilicals. All sealines are anchored to the structures.

Fig. 2 shows the principle arrangement of Manifold Station "A".

## LINES AND UMBILICALS

The following interfield flowlines and umbilicals connect the Lille-Frigg Manifold Station "A" to Frigg TCP2 platform:
- 10 inch rigid steel gas production pipeline
- Flexible bundle containing:
  - 3 x 1.5 inch chemical injection lines (one for each well)
  - 3 inch service line
- Hydraulic umbilical
- Electrical umbilical

These lines are trenched and back filled for protection apart from the platform approach, manifold station approach and crossings of existing pipelines. Untrenched sections are protected by concrete mattresses or rockdumping.

The following infield flowlines and umbilicals connect Manifold Station "A" to each of the satellite stations "B" & "C":

- 4 inch flexible gas production line
- 1.5 inch chemical injection line

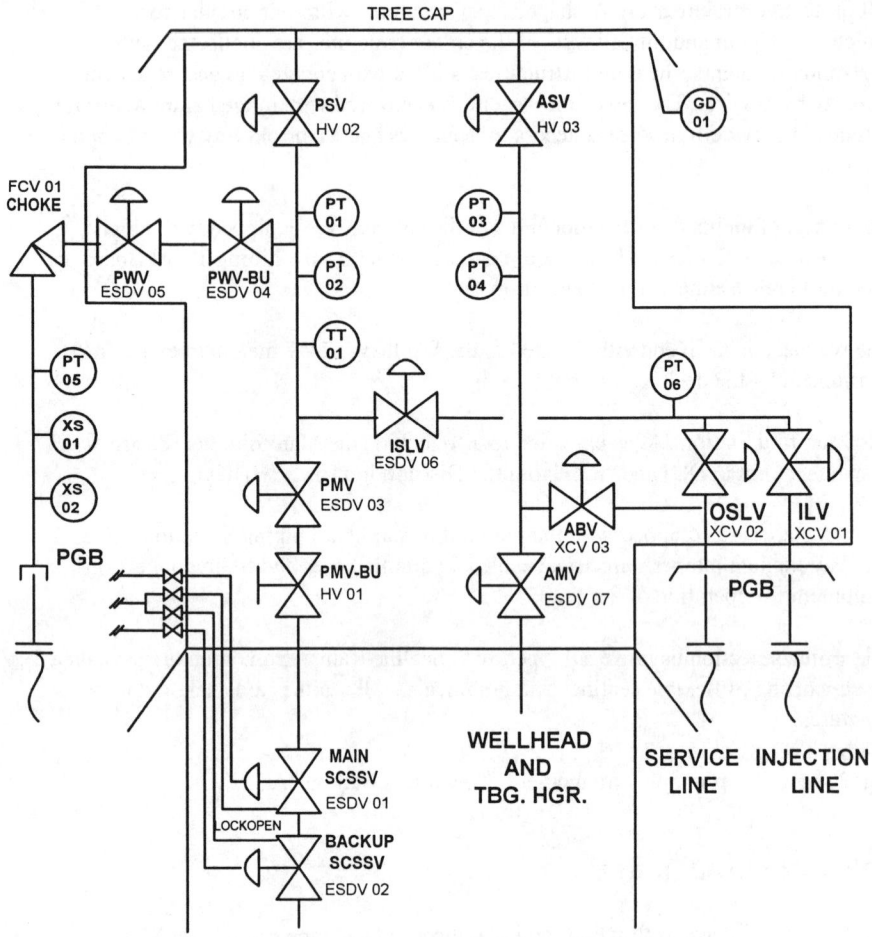

Fig. 3
## LILLE FRIGG SUBSEA
## XMAS TREE LAYOUT

- 1.5 inch service line
- Hydraulic umbilical
- Signal and power umbilical

Interconnections within the protective structures are made by divers.


## SUBSEA PRODUCTION SYSTEM

### Wellhead System

The wellhead system consists of a 30" conductor housing, 18 3/4" high pressure wellhead housing, 13 3/8" and 10 3/4" casing hangers with associated pack-off seal assemblies.

The system interfaces with Production Guide Base, Temporary Guiding device, Tubing Hanger, the X-mas tree, and supports the well casings.

The system facilitates the drilling and completion of the well and subsequent production.

The wellhead system is rated at 15000 psi with an expected maximum wellhead shut-in pressure of 7800 psi.

### X-mas Tree System

The X-mas Tree is dual bore, 4 1/16" production and 2 1/16" annulus, rated for 10000 psi.

The main components of the X-mas Tree Assembly are as follows:

- 18 3/4", 15M Wellhead connector
- X-mas Tree Valve Block
- Valves and Actuators (5 off 4 1/16", 6 off 2 1/16")
- X-mas Tree Mandrel
- Guide and Protection Structure
- Production Flowloop
- Choke (retrievable insert/actuator design)
- Service/Injection Flowline
- Monitoring Equipment
- Mounting Base for Well Control Pod
- Hydraulic piping between Mounting Base and Actuators

The production choke insert and Well Control Pod are installed on the X-mas Tree, and are independently replaceable from the X-mas Tree by use of running tools and ROV assistance. The X-mas Tree is suitable for diverless, guideline and ROV assisted installation and retrieval, using dedicated tools run by workover riser or drill string. Fig. 3 shows the general arrangement of the X-mas Tree.

**Manifold Assembly**

The Manifold Assembly has been designed to provide the following operations:

- To be retrieved and re-installed independently from Manifold Station "A"
- Allows vertical connection of a subsea pig launcher to provide a subsea pigging capability
- Allows vertical connection of a workover riser to facilitate a well kill operation on any well as required
- Isolate up to five infield production flowlines
- Isolate up to five infield service flowlines
- Allows cross-over facilities
- Manifold up to five infield production flowlines into one interfield production pipeline
- Includes ROV operable valves and connector overrides
- Includes flange connections to allow maintenance using diver intervention

## SUBSEA CONTROL SYSTEM

The subsea control system constitutes a vital part of the overall subsea production equipment, and its reliability is paramount in order to achieve safe and undisturbed production.

The main challenges and important implications for the LF subsea control system design have been strongly influenced by the high reservoir pressure and temperature conditions.

The use of standard downhole safety valves (SCSSV) technology would normally require a very high pressure (VHP) control supply, usually about 2000 psi above the shut-in wellhead pressure.

During the concept evaluation phase of the LF development, several manufacturers of SCSSVs were approached in order to evaluate the required control fluid operating pressure. A nominal 11000 psi WP (including safety margins) was defined based on manufacturer's recommendations.

The SCSSVs and their associated control lines filled with a column of stagnant control fluid are the only elements of the control system affected by the high pressure and temperature, thus defining the design constraint for other hydraulic components. In 1989/90, field proven commercially available hydraulic components such as control fluid, control valves and couplers able to satisfy the stated pressure and subsea application requirements were not available. This situation initiated an Elf in-house qualification programme for such components.

Another important hydraulic component essential for long distance subsea control is the VHP hoses used in the umbilicals and jumpers. The manufacturer of the LF umbilicals initiated early 1991 a qualification programme of hoses designed for 11250 psi WP. These tests included burst pressure tests at minimum 45000 psi and impulse testing at 15000 psi for 200000 cycles.

Based on results from development programmes and qualification tests performed by Elf in France and Norway on subsea mateable conductive electrical connectors and prototype ROV tooling, the LF concept has enhanced the design and development of a specific ROV operated coupler tool. This is a combined tool for connection/disconnection of both electrical and hydraulic couplers, incorporating a low torque tool.

## Qualification Program

Prior to the development phase, the following component qualification tests were undertaken by Elf:

Hydraulic fluid

Two types of control fluids (mineral/synthetic oil) were short-listed for qualification testing.

Fluid samples of each type were initially submitted to 150°C and 200 bar environmental conditions over a 21 days period. No significant deterioration was experienced when analyzing the physical characteristics and chemical compositions after completing the test.

Thereafter, more fluid samples were submitted to 120°C and 770 bar environmental conditions over a period of 140 days. The test bench set-up procedures did not include particular precautions with regard to trapped oxygen and humidity in the test cells.

After 24 days of testing, aged fluid samples were submitted to laboratory analysis similar to the initial test. Strong degradation of fluid properties (solid particles, colour change) caused by oxidation and hydrolysis processes was revealed. It was also possible to determine that these degradations were triggered by some specific additives like the antiwear additive.

Alternative ways of substituting the concerned additives were investigated, including possible impacts on the components of the control system.

Samples of new formulated fluids (2 types) were then submitted to similar series of high temperature and pressure aging tests as the previous fluids. However, these new fluids also had small amount of trapped air and seawater included. The set of laboratory analysis performed on the aged samples confirmed that the chemical stability had been substantially improved and would be acceptable for LF application.

Hydraulic couplers

Earlier research activities led by Elf's Research Centre in France had resulted in development of a new self sealing, pressure balanced, very high pressure rated hydraulic coupler. Prototypes had been built, tested at 10000 psi in the laboratory and in the shallow water test site of Bayonne, France. Extended shallow water tests were subsequently initiated to qualify an improved version of these couplers at 11000 psi, using the control fluid selected for LF application. Three stab type couplers were installed on a remotely operated test bench, immersed in turbid seawater, pressurized to 11000 psi and successfully submitted to long term endurance tests with periodic connection/disconnection cycles immersed.

Solenoid operated control valves

In July 1990, a market survey was performed amongst potential control valves manufacturers who had previously demonstrated a capability to deliver such valves for subsea remote control systems. Two manufacturers were short-listed, both willing and capable of supplying pilot operated solenoid driven shear seal valves designed for operating at 11000 psi WP.

Four valves of each type were tested according to the following programme:

| | |
|---|---|
| Phase 1 | - Acceptance test |
| | - Leakage test |
| | - Proof pressure tests |
| | - Functional tests under operational temperature conditions |
| | - Shocks and vibration tests |
| | |
| Phase 2 | - Wear and tear test |
| | 5000 cycles under operational pressure and temperature conditions |
| | |
| Phase 3 | - Long term test |
| | 6 months under conditions as close as possible to real operational conditions |
| | |
| Phase 4 | - Leakage tests |
| | - Proof pressure tests |
| | - Functional tests under operational temperature conditions |

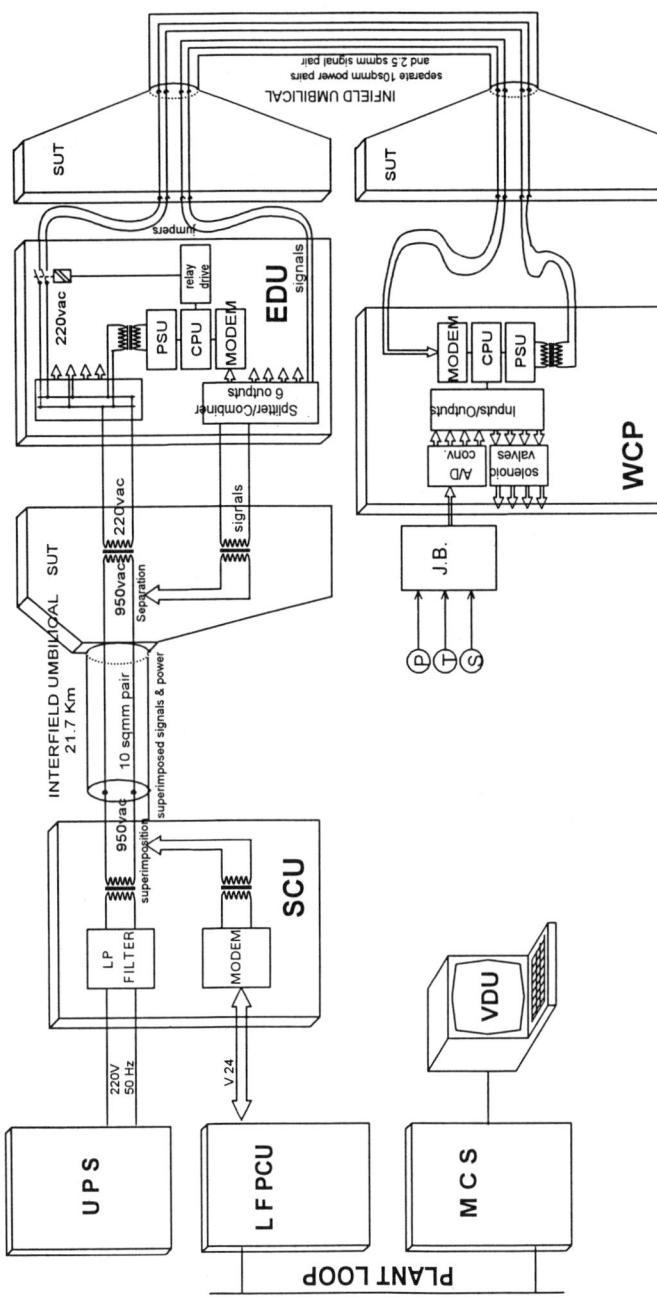

LILLE FRIGG SUBSEA
Fig. 4  TRANSMISSION SYSTEM DIAGRAM

## ELECTRICAL AND ELECTRONIC SUBSYSTEM

The electrical/electronic susbsystem (see Fig. 4) consists of the uninterruptable electrical power supply units, the Subsea Communication Unit (SCU) and the Computer Control System, all located on the TCP2 platform.  These units supply electrical power and provide two way communication signals to the subsea Electrical Dispatcher Unit (EDU) via the interfield electrical umbilical and the individual WCPs via the infield electrical umbilicals to facilitate the following tasks:

- operate solenoid driven hydraulic control valves located in the WCPs
- operate process date acquisition systems located in the WCPs
- operate power isolation relay switches located in the EDU
- operate a power distribution data acquisition system located in the EDU

### Surface equipment

The surface equipment includes:
- An Electric Power Unit which provides the Lille-Frigg subsea equipment with required electrical power.  This unit consists of a redundant uninterruptable power supply (rectifier, battery and converter).  Two separate 220 V, 50 Hz distribution cables supply electrical power to the SCU.
- The SCU provides the main interface between the platform uninterruptable power supply, the Lille-Frigg Process Control Unit (LF-PCU) and the interfield electrical umbilical, and superimposes the communication signals onto the power line for subsea transmission.
  The SCU incorporates dual redundant 220 V to 950 V 50 Hz step-up transformers and dual redundant communication modems and line interface units.
- A Computer Control system which is installed for surveillance, interaction by operators and automatic control of the subsea installations, including:

  * a Processing and Control Unit (PCU); provided with fully redundant subassemblies including power supply modules, several task-dedicated multifunction processors (microprocessors), input/output modules and communication adaptors.
  * a number of Management Command System  units (MCS), each allowing the graphic presentation of the subsea installations status and alarm modes, and providing access for operator control interaction.

The Processing and Control Unit operates as an independent controller for all subsea electronic units (EDU,WCPs), including data processing for status and alarm presentation, control logic of subsea process and emergency shutdown function, communication protocol and subsea units dialogue surveillance.

In addition, the Processing and Control Unit exchanges data via a plant loop (data high way) with the Management Command System and other surface controllers (controlling chemical injection units, hydraulic power unit, electrical power unit and gas/condensate process) as well as hardwired I/O signals interconnected with Frigg ESD system. Although specifically dedicated to the subsea production stations control, the Processing and Control unit is an integral, distributed controller of the Frigg Control and Data Acquisition (FCDA) system, and derived from a manufacturer standard product range.

Amongst other advantages, this allows the operator to coordinate and execute supervisory and control tasks of both subsea and surface installations from the same operator stations (MCS units).

The Processing and Control units execute major processing tasks which have been implemented due to the inaccessibility of subsea components for test and repair, e.g. logic reconfiguration, inhibition of data feedback, automatic redundancy changeover etc. This will facilitate elimination of control function effects and nuisance to operator by single subsea component failure.

## Electrical umbilicals

The electrical link between surface equipment and the subsea production stations ensures a dual redundant configuration for transmission of power and signals. This is achieved by the Interfield electrical umbilical (21,4 km long).

The electrical interfield umbilical assembly comprises the following main elements in the installed conditions:

- topside umbilical splice making connection to platform wiring from the Electrical Power unit and the Subsea Communication unit (SCU)
- umbilical final manufacture in single length
- individual twisted pair cables manufactured in two lengths with mid length splice made during lay-up
- subsea umbilical termination(SUT) incorporating step-down transformers (950V/220V), filter network, matching transformer and weak link

The two off screened twisted pairs are providing dual redundant power and signal (power transmitted between individual conductors, signal transmitted between both conductors and screen).

The cross-section of the interfield umbilical is shown in Fig.5. Infield electrical umbilicals link Manifold Station "A" with the two other Satellite Stations. These umbilicals are of a similar construction as the interfield umbilical, however differing by having the polypropylene filler ropes replaced by two off screened twisted signal pairs, equipped with SUTs in either end, and with shorter lengths of 1 km and 1,4 km respectively.

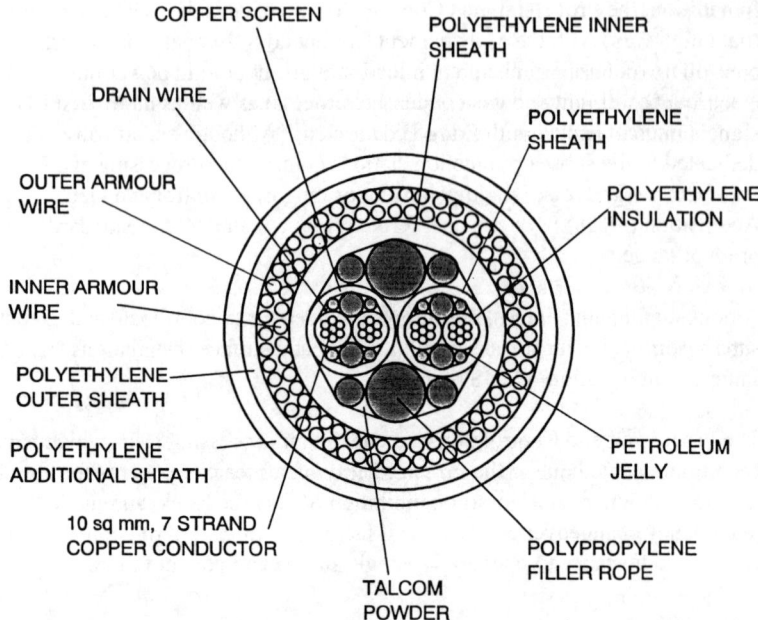

COPPER SCREEN

POLYETHYLENE INNER
SHEATH

DRAIN WIRE

POLYETHYLENE
SHEATH

OUTER ARMOUR
WIRE

POLYETHYLENE
INSULATION

INNER ARMOUR
WIRE

POLYETHYLENE
OUTER SHEATH

POLYETHYLENE
ADDITIONAL SHEATH

PETROLEUM
JELLY

10 sq mm, 7 STRAND
COPPER CONDUCTOR

POLYPROPYLENE
FILLER ROPE

TALCOM
POWDER

Nominal O.D.   : 65 mm

LILLE FRIGG SUBSEA
Fig. 5  INTERFIELD ELECTRICAL UMBILICAL

**Electrical Dispatcher Unit (EDU)**

The Electrical Dispatcher unit installed on Production Station "A" is incorporated to facilitate distribution of power from the surface power supply unit to the subsea loads and dispatch signals to and from the subsea control units (EDU itself and WCPs).

The electronic sub-assembly of the EDU is made up of five fully dual redundant functional units comprising:

- communication interfaces
- data acquisition and control system
- power splitting
- signal splitting
- internal power supplies

The first sub-assembly (communication interfaces) is exactly identical to the sub-assembly used within the WCP for the same function. However, the remaining four sub-assemblies are unique for the EDU.

- The communication interface allows the EDU to transmit and receive signals in FSK modulation and in half duplex mode. A standard modem is used at 1200 bps. The modem interfaces satisfy V24 and V23 CCITT requirements for user (microprocessor) and line interfaces respectively.
- Data acquisition and control system are microprocessor driven. The architecture of the system is a central processing to interface PC boards connected on a dedicated bus. All these interface PC boards are galvanically insulated from the CPU which can directly receive all internal monitoring parameters.
  The power received from the interfield umbilical (220V, 50Hz) is split in six directions. Five power outputs are routed to each of the WCPs (including 2 spare), and one power output is used for internal EDU power requirements. The distribution/splitting of the incoming power supply is performed using an internal cable matrix.

  Each WCP output power line has an integrated dual state relay, which is used to interrupt power transmission to the WCP when commanded by the surface controller.
- The communication data signals from the interfield umbilical are also split into six directions, one internal and five external. This is achieved using a passive splitter/combiner, consisting of a transformer with one primary and six secondaries.
- The EDU receives two separate power supply inputs (220V, 50Hz) from the interfield umbilical jumper cable connections.

  The input power supply line is connected to a step down transformer which delivers two low voltages on the secondaries. The two low voltage AC secondary circuits are then linked to serial regulators on dedicated PC boards.

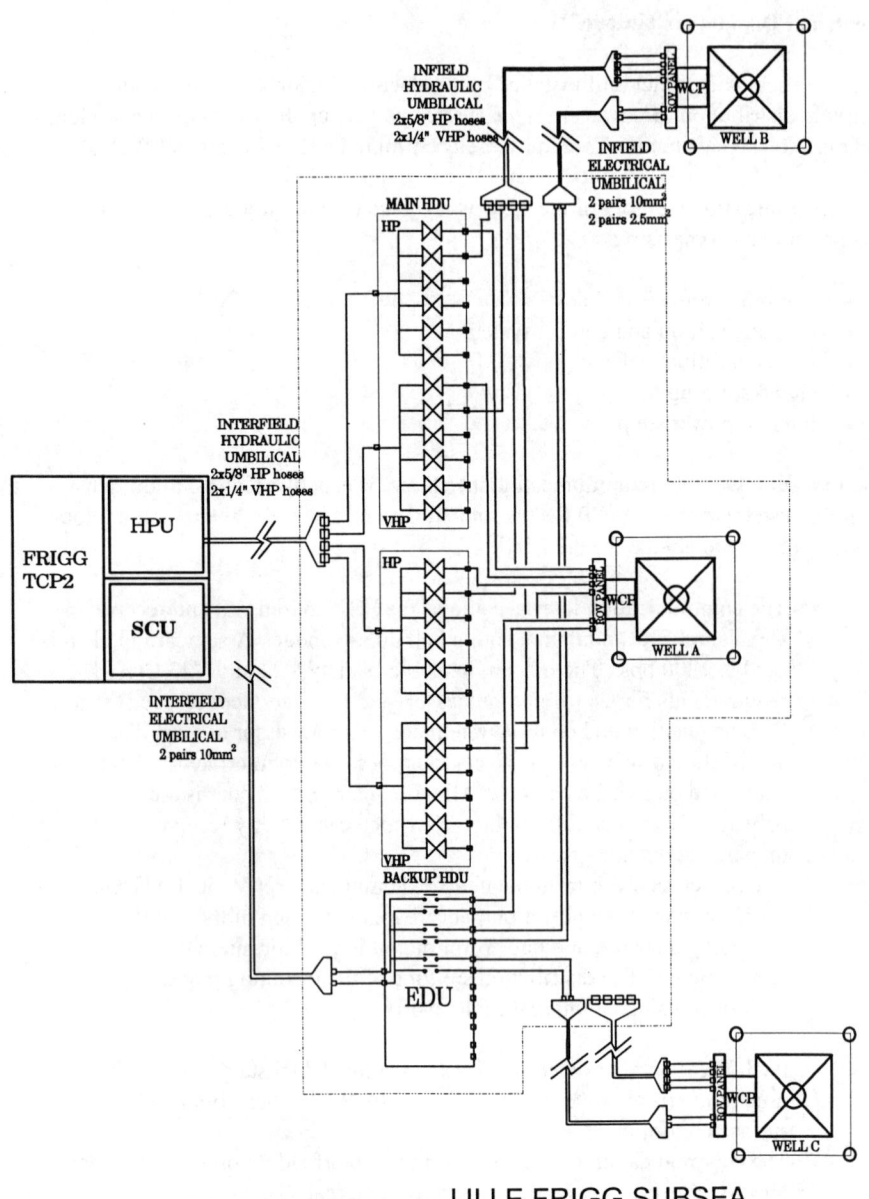

LILLE FRIGG SUBSEA
Fig. 6    REMOTE CONTROL SYSTEM

Two DC voltages are also required to drive the EDU electronics:
+ 15V DC for the Central Processing Unit (CPU) and the modem
+ 24V DC for the control of the switching relays

A cross strapping PCB (power supply redundancy) is installed in order to allow for cross redundancy at all power levels (CPU and Relay drive).

## Signal Transmission System

The LF signal transmission system can be split into 3 main parts (See Fig. 4):

- The surface related signaling equipment and the 21,5 km long interfield electrical umbilical which contains 2 screened twisted pairs (main/back-up) and terminated at the subsea end with a SUT.

- The EDU which receives the communication signals (main/back-up) and splits the signals into 6 directions or combines the signals from 6 input lines to 1 output line.

- Redundant communication lines (main/back-up) which are connected to the EDU at one end and to each of the modems of the WCPs via a network of electrical jumper cables and infield electrical umbilicals at the other end.

The signal transmission system is particular in the way that signals are superimposed on power between Frigg and the subsea termination head of the interfield umbilical installed on Production Station "A".

Separation of the signal path from the power line is taking place in the interfield SUT. Subsequent subsea dispatching of signals is achieved in the EDU by means of passive components (splitter/combiner circuit). The splitter/combiner is designed to allow up to five WCPs to be connected in addition to the modem of the EDU itself.

Signals are transmitted in half-duplex mode using Frequency Shift Keying (FSK) at 1200 baud between the Master CPU (LF PCU) on the platform and each of the 4 subsea Slave CPUs within the EDU and WCPs. Only the Master CPU is allowed to initiate communication.

## Hydraulic Subsystem

The hydraulic subsystem (See Fig. 6) consists of a dedicated hydraulic power generation unit (HPU) installed on Frigg TCP2 platform, an interfield umbilical linking the surface HPU and Manifold Station "A", hydraulic distribution units (HDUs), jumper hoses, infield umbilicals linking the hydraulic network on Manifold Station "A" with the satellite stations and finally jumper hoses linking the umbilical terminations with the Well Control Pods (WCPs). In the WCP the hydraulic fluid is distributed and controlled to open/close X-mas Tree valves and SCSSVs.

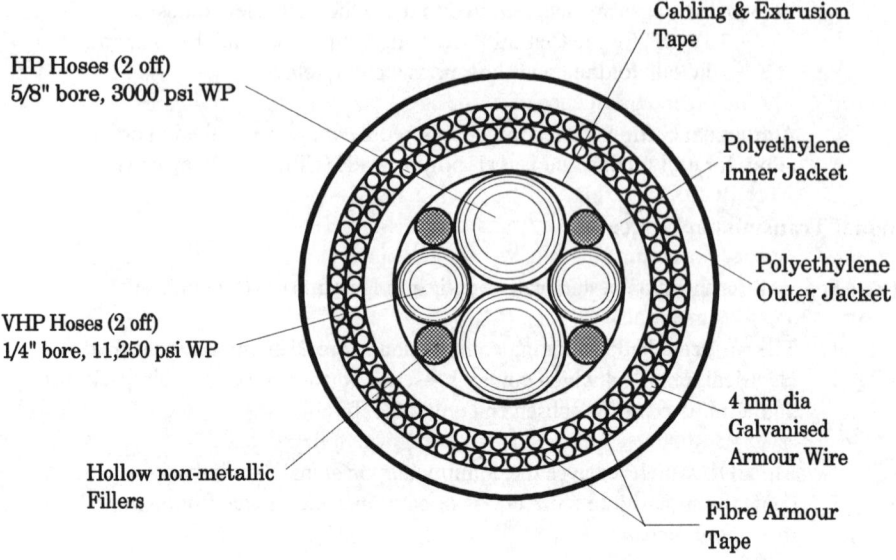

HP Hoses (2 off)
5/8" bore, 3000 psi WP

Cabling & Extrusion
Tape

Polyethylene
Inner Jacket

Polyethylene
Outer Jacket

VHP Hoses (2 off)
1/4" bore, 11,250 psi WP

4 mm dia
Galvanised
Armour Wire

Hollow non-metallic
Fillers

Fibre Armour
Tape

Nominal O.D.   :   85 mm

|  |  |
|---|---|
| **Fig. 7** | **LILLE FRIGG SUBSEA** **HYDRAULIC UMBILICAL** |

The main features of the subsystem can be summarized as outlined below:

- The open loop hydraulic system discharges hydraulic fluid to sea, primarily during opening sequence of X-mas tree valves and to a limited extent during the closing sequence.
- The hydraulic fluid used is Subhydrelf 15NF mineral oil type, which has been qualified for high pressure and temperature application. A similar fluid has previously been applied by Elf for the North East & East Frigg subsea developments, selected for its good behavior and low viscosity at low temperature as well as good knowledge of its compatibility with various materials.
- The dedicated HPU installed on the surface delivers two levels of hydraulic pressure:
  VHP: 11000 psi Working Pressure (WP) for all SCSSV functions
  HP:     3000 psi WP for all X-mas tree valve functions
  It incorporates the necessary pumps, filtration, instrumentation, accumulators and regulation circuitry to permanently feed the subsea hydraulic network at required pressure level and cleanliness (class 6 of NAS 1638) in order to allow adequate control of subsea valves from the Management Control System at Frigg.
- The interfield umbilical from Frigg TCP2 platform (21.4 km long) supplies hydraulic pressure from the platform HPU to two HDUs located on the Manifold Station "A". It incorporates four hoses as shown in Fig. 7:
  - 2 off VHP supply hoses, (one for main supply and one for back-up supply) 1/4" bore, 11250 psi WP thermoplastic hoses.
  - 2 off HP supply hoses (one for main supply and one for back-up supply) 5/8" bore, 3000 psi WP thermoplastic hoses.
- Each of the two HDUs distributes HP and VHP hydraulic supply to the Well Control Pods associated with X-mas trees "A", "B" and "C". The HDUs incorporate ROV operated isolation valves on the HP and VHP lines to enable depressurization and removal of the individual Well Control Pods. The HDUs are linked to the WCP in Manifold Station "A" by separate HP and VHP pre-installed and diver connected jumpers, incorporating balanced stab hydraulic couplers. Likewise, the HDUs are linked to the WCPs on Station "B" and "C" by separate HP and VHP preinstalled diver connected jumpers and infield umbilicals.
- The infield umbilicals between Manifold station "A" and Satellite Stations "B" (1 km long) and "C" (1,4 km long) have an identical construction as the interfield umbilical.
- The final hydraulic control elements in the WCP are described separately below.

**Well Control Pods (WCPs)**

General

The WCPs are identical and interchangeable assemblies that can be remotely run and retrieved from the subsea X-mas tree using a purpose built running tool in combination with ROV intervention tooling or divers.

Each WCP is located on the X-mas tree using the well control pod mounting base (WCP MB). The WCP to WCP MB interface incorporates wet mateable electrical and hydraulic couplers, for controlling and monitoring the X-mas tree hydraulic functions and process sensors.

The WCP is hooked up to the subsea hydraulic and electrical distribution systems by electrical and hydraulic supply jumpers. The jumpers are dual redundant assemblies that are connected to the WCP using ROV intervention tooling or divers.

The WCP incorporates hydraulic and electronic sub-assemblies for controlling the X-mas tree functions and monitoring the input from X-mas tree mounted process sensors. The electronic sub-assembly incorporates dual redundant control and partial redundant data acquisition circuitry.

Mechanical arrangement and interfaces

The WCP comprises a central main plate with hydraulic and electronic sub-assemblies bolted to its upper and lower sides and an external guide frame assembly that bolts to its outer edges.

The main plate houses all the external electrical and hydraulic couplings that interface with the X-mas tree and supply circuits. The X-mas tree couplings are located on the lower-most outer edge of the main plate to provide an automatic make-up of the couplers during installation of the WCP onto the WCP MB. To aid the make-up of the couplers, the main plate is fitted with fine alignment dowels that make up with the WCP MB.

The supply couplers are located on the outward facing side of the main plate behind the ROV panel.

The electronic outer container is used as the primary method of aligning the WCP to the WCP MB during installation, and incorporates longitudinal ribbing for guidance.

The hydraulic sub-assembly comprises of a protective outer can, a protective can mounted running tool neck and sub-plate mounted hydraulic control valves and associated components.

Electronic Sub-Assembly

The WCP electronic sub-assembly comprises :

- a control and data acquisition system
- a communication interface unit
- power supply units that provide the different voltages required to drive the electronics.

The control and data acquisition system is a fully redundant system that comprises two parallel and inter-linked control and data acquisition units for controlling and monitoring the X-mas tree valves, choke valve, downhole SCSSVs, process data and the WCPs internal operating conditions.

The two control and data acquisition systems (main and back-up) incorporate the following principle components:

- microprocessor
- digital acquisition interface board
- solenoid valve driver interface board
- individual solenoids
- dedicated bus
- pulse transformers

The main and back-up control and data acquisition system are principally the same but differ in that the back-up unit is only provided with essential data acquisition circuitry.

The solenoid driver interface board is fitted with fuses on every drive function output. The fuses are designed to be blown in the event of a short circuit on the valve driver board maintaining the solenoid in the energized position.

The communication interface unit is a fully redundant system that comprises two communication modems interlinked by a modem/processor interface. The modems transmit and receive communication signals with the platform control system using FSK modulation and in half duplex mode.
The modems satisfy CCITT V24 and V23 recommendations for data transmission and line interfaces. The modem transfers data at 1200 baud.

Two identical power supplies provide the various voltages required to operate the WCP electronic sub-assemblies. Each of the power supplies is capable of operating the WCP independently of the other.

Hydraulic Sub-Assembly

The hydraulic sub-assembly is made up of a main plate which is used as a mechanical base for the hydraulic components and as an interface for the WCPs supply and output function connections.

The main plate is used for mounting the following hydraulic components that make up the WCP;

- isolation solenoid operated valves
- monostable solenoid operated valves
- bi-stable solenoid operated valves
- switch over solenoid operated valves
- monostable hydraulic operated valves
- VHP pressure transducers
- HP pressure transducers
- HP pressure switches
- flowmeters
- manifold bleed valves
- ROV hydraulic supply couplers
- WCP/WCP MB hydraulic couplers
- HP hydraulic accumulator
- ROV operated bleed valves

The control valves are shear seal type valves which are mounted to the main plate by means of sub-plates. The sub-plates incorporate outlet filters, filter bypass valves and pressure sensors.

The hydraulic assemblies electrical components are connected to the electronic assembly via the main plate using multi-pin electrical penetrators and inter-connecting dielectric oil filled wiring harness.

The electrical penetrators are secured to the main plate and provide a pressure containing barrier between the hydraulic and electronic enclosures.

The inter-connecting wiring harness clamps to the hydraulic components and electrical penetrators providing a secondary protective barrier to the enclosed cables and end connections.

The WCP has the following hydraulic design requirements:

- HP working pressure   :        207 bar (3000 psi)
- VHP working pressure :        759 bar (11000 psi)
- hydraulic filters         :        10 microns absolute
- control fluid             :        Subhydrelf 15NF
- fluid cleanliness       :        NAS 1638 class 6

## Running Tool

Installation/retrieval of the WCP, HDU and EDU is achieved by use of a dedicated running tool which can be attached to the top of the units by means of an automatic latching mechanism. Unlatching can be performed either by the ROV Intervention tooling or by divers.

The running tool is operated on two guidewires and equipped with soft landing cylinders, ROV docking facilities and removable guide post funnels to match the interface constraints of the individual units.

## ROV Intervention Tooling

The ROV intervention tooling adapted for LF applications is a field proven tool deployment system operated in conjunction with positive docking into predefined intervention points on the control system units and running tool.

The tool deployment system is a 3-axis Cartesian device configured to suit the operational requirements.

A special connection tool has been developed for attachments to the deployment system and provide interface with both the electrical and hydraulic coupler halves of the supply jumpers attached to the WCP ROV panel. The coupler tool also incorporates a low-torque tool required to operate the WCP lock-down screws and bleed valves. Thus the ROV is able to perform all required tasks for a WCP replacement in one single dive. This tool is moreover used for operating the isolation valves on the HDUs. The tool deployment system is prepared to perform operations on the production related equipment (X-mas tree, Manifold, Workover tools) by replacing the coupler tool with other torque tools. A topside control console located on the support/diving vessel is used to control and monitor the intervention tasks via spare cables in the ROV umbilical.

## CONCLUSION

The Lille Frigg field development is incorporating a remote control system particularly designed to meet hostile operational conditions. A new control pod concept has been applied in combination with components specifically qualified to satisfy demanding temperature and pressure conditions.

Several technical challenges had to be addressed, tested and solved within a rather limited time frame.

Although the concept adopted is requiring a combination of diver and ROV assistance during installation and intervention, it is feasible to adapt the concept to become fully diverless.

Most of the system has already been installed, wells are being completed and commissioning activities planned towards the end of 1993.

# A MOVE TOWARDS SUBSEA CONTROL SYSTEM STANDARDISATION?

Roy Windsor
GEC-Marconi Oil & Gas
2 High Street, Nailsea, Bristol
and
Per Arne Nilsen
Norsk Hydro
PO Box 190, N-1321 Stabekk, Norway

## ABSTRACT

System standardisation is acknowledged to be one area which can significantly reduce the life cycle costs for any project. This paper looks at subsea control systems, specifically the Norsk Hydro Troll Olje Project in the light of previously defined objectives.

## INTRODUCTION

The subject of standardisation for subsea control systems has been broached several times. This review takes as its starting point the paper presented at Subsea Control and Data Acquisition (1992) titled "Standardisation of Subsea Control Systems". Since that time, the first deliveries to the Troll Olje project have taken place and we are in a position to evaluate some of the theories and practicalities in the light of a "real life" project.

The alternatives of a "Company Standard" or "Vendor Standard" both have drawbacks, however the use of commercially available products which have been pre-qualified seems to be a path that several vendors and oil companies might like to consider.

Since the system suppliers are in continual competition in the world market, the prices should be market driven and a monopoly situation is less likely to occur. The "market" is governed by oil companies and they have different views and requirements which may make standardisation more difficult to achieve.

We will show the areas where "standardisation" is reasonably easy to achieve and look at the more difficult topics such as the Subsea Electronic Module (SEM) where the challenge is greater.

*Volume 32: Subsea Control and Data Acquisition, 27–40.*
© 1994 *Society for Underwater Technology. Printed in the Netherlands.*

**The Starting Point for Standardisation**

In late 1991, Norsk Hydro started presenting their views to various subsea control vendors with the intention of having some "standardisation" in place when the Troll Olje subsea control package came out for tender.

There were some general function requirements and some specific equipment requirements:-

•       an electrohydraulic system was desirable;

•       subsea communication would be superimposed on the power cables;

•       it would be possible to modify subsea software from topside;

•       subsea software was to be written in a high level language;

•       Subsea Control Unit (SCU) would be an integral part of the DISCOS (Distributed Supervision, Control and Safety) system;

•       Hydraulic Power Unit (HPU) would be part of the normal topside equipment;

•       Emergency Shutdown (ESD) would be obtained by means of a system physically separated and different from the Process Shutdown (PSD) system;

•       Xmas tree instrumentation would not influence the workover frequency;

•       Interfaces should be standardised and resolved by the various subcontractors.

With these requirements as a base, and a range of pre-qualified components, the subsea control vendors were asked to quote against a well tried solution using "standard" components.

Norsk Hydro had all the experience of TOGI and previous projects to hand when reviewing the various solutions and components, and were able to ensure that known pitfalls were avoided. The GEC-Marconi Oil & Gas (G-MOG) solution was to fit the pre-qualified components into an existing and tried solution which has evolved over many projects.

"A move towards subsea control system standardisation?" This is the question we will now attempt to answer.

**The Realisation**

Once the contracts had been awarded, it became obvious that the interfaces would need a lot of attention and that standardisation of mechanical interfaces would be more difficult to achieve than the electrical and communication interfaces.

Appendix A - Troll Olje Production Control System Overview - is intended to give some background information about the practical solution which forms the basis for this standardisation assessment.

The world market for subsea control systems is not sufficient to enable volume production and therefore, in addition to the use of commercially available components, a standard approach has to be applied for each supplier to the lockdown mechanism and manifolding (see Figure 2 and 3).

The main components chosen were:-

| | | | |
|---|---|---|---|
| • | Flowmeters: | EG&G Flow Technology (Phoenix) | USA |
| • | Directional Control Valves: | Rotator | Norway |
| • | Electrical Connectors: | Lockheed Marine | USA |
| • | Tree Sensors: | Read Matre Instruments | Norway |
| • | Hydraulic Couplers: | National Couplers | USA |
| • | Lockdown Mechanism: | Costain-Fuel | UK |
| • | SEM: | GEC | UK |
| • | Internal Electrical Penetrators: | Brantner/Seacon | USA/UK |
| • | Communication Interface Protocol: | TC 57 on Hughes Aircraft Corporation Modem (IEC-870-5) | |

The SEM is a new design based on the 68000 series μP. For all new cards, the VME bus interface has been used, but for the existing card designs GEC's proprietary SEMIBUS has been retained.

This makes Norsk Hydro dependent on GEC for all work related to the SEM.

Such dependencies may continue to be acceptable, provided that the supplier is expected to be in the market in the future, the availability of necessary resources can be guaranteed and that the services are costed according to market level.

Whether or not to take the next step and go to a full VME system based on standard size VME cards available on the open market is very dependent on the above, and indeed the cost (CAPEX and OPEX) for a fully VME-based system.

Here is perhaps where the next step should be taken.

The criteria for component selection has been "accepted standards", in this case standard mechanical fittings such as Swagelock or Autoclave fittings, and standard electrical signals, e.g. 4-20mA. Although this satisfies many of the standardisation objectives, we are still faced with the problems of single source and obsolescence. The question is "What happens if a component is no longer manufactured?" or even worse for the oil companies "What if the main contractor ceases to exist?".

New mechanical components can be contained within the same space envelope and mounted on conversion pieces. These are also electrically not difficult to obtain to industrial standards, e.g. 24 VDC, 4-20mA, RS232, etc. The exception is the tree sensor interface which has a unique output signal.

A component, such as the Retlock central locking mechanism, can be manufactured to drawing if the supplier goes out of business. This, in fact, applies to all the mechanical components.

Replacement of the SEM is also possible but here there are restraints on maximum physical size and electrical connections which have to make use of existing penetrator components.

Software has been written in C++, which is fully downloadable to the subsea control pod, allows great flexibility, and the use of recognised and commercially available protocol means that this is an area where changes can be readily implemented.

The chosen design for the Control Pod Mounting Base and the Running Tool are not easy to standardise. The X-tree to Control Pod Mounting Base interface is defined by the dimensions of the two mechanical structures. Since we have control pods fitted to all the major types of X-tree, our flexible solution results in minimal adjustments. The physical constraints of the manifold and X-tree may also dictate that a "standard" design cannot be utilised.

Our solution for a central lockdown mechanism may not completely meet the Norsk Hydro requirement of only having one Running Tool for several suppliers' control pods but, maybe a common Running Tool "frame" with smaller "attachments" which would accommodate any differences for future development projects will be the solution.

## CONCLUSION

The general guidelines for the use of pre-qualified, proven, and commercially available components, did not inhibit our design but restricted the number of vendors able to tender.

For the oil companies, the most important drive for standardisation is the reduction of life cycle cost (OPEX). In addition, reduced engineering activities, project execution times, testing and qualification will reduce the initial investment cost (CAPEX).

On the Troll Olje project some requirements for standardisation were included in the contracts, but as described in the referenced paper, not everything can be implemented at the same time in one project. There is also the question of gaining experience and making sure that the implementation of standard system/components fulfils the goals.

Specifically what was achieved in Troll Olje:-

• As described, many component vendors were short listed based on experience from previous projects now in production. This ensured that quality components were selected, and increased the probability for improved lifetime reliability. In addition, the selected vendors have been on the market for many years, with no indications that any should disappear form the market in the foreseeable future.

- Electrical interfaces to the instrumentation in the control pod have been made using standard format/signal. This makes it easy to replace instruments from one vendor with another, and avoids the need to change the mechanical design of the pod. There is also no need to redesign the SEM interface cards.

- The use of a company specified communication protocol, based on an international standard (IEC-870-5), has made it easier for companies' operational personnel to familiarise themselves with the protocol, and adjustments/changes during the project execution have been possible based on operational experience from previous projects. In addition, this will ease the test and operation phase if fault tracing should be necessary.

- ROV tooling, such as valve tool and torque tool, from the previous project (TOGI) was specified to be used. This meant that all ROV interfaces have been designed to meet the specification for the TOGI tooling. Thus, the CAPEX and OPEX for new ROV tooling was saved.

What would have been desirable to implement, but could not be included on the Troll Olje project?:-

- Use of already existing control pod running tool. This was not possible since the interface and locking to the Xmas tree on the previous project was a proprietary design.

- Use of standard signal format for the electrical interface. Although the same vendor for process sensors as for TOGI was selected on Troll Olje, the interface (electrical) is changed. Consequently, the interface card for the sensor in the SEM had to be redesigned. The electrical interface here is not on a standard signal format.

- Use of a standardised SEM. The SEM is still considered to be a proprietary unit. Although it is much more "open" than previous systems, maintenance can only be performed by GEC. The changes from previous projects are:-

  - All software is programmed in a high level language (C).

  - Norsk Hydro have requested all software documentation and sufficient hardware information to perform fault tracing to board level.

Although Troll Olje has taken some small steps closer to the concept of standardisation on control systems, the important interfaces where significant cost savings might be achievable have not been standardised. This applies to the interface between the control pod and Xmas tree and the control pod to running tool.

Standardisation is possible provided the working relationship between the oil companies and the suppliers can be developed to a point where all parties have the same objectives.

# APPENDIX A

# TROLL OLJE
# PRODUCTION CONTROL SYSTEM
# OVERVIEW

## 1.0    Introduction

The Troll project consists of a large subsea development taking place around two separate provinces. The first stage is the development of the oil province which consists of four cluster manifold stations, each connected to the Floating Production Unit (FPU) by an integrated service umbilical. Each manifold is designed for six well tie-ins although only 19 wells are planned for the whole province.

The second stage will be the development of the gas province which will consist of five cluster manifold stations. Six station manifolds will again be used except in one case where an eight station manifold will be employed, providing a maximum capacity of 32 wells overall. The development plan currently includes 25 wells to be provided across this province. Figure 1 outlines the projected field layout.

The minimum offset distance between FPU and manifold is approximately 4.3 km, while the maximum design offset is 25 km. The individual wells will be situated between 50 m and 3.6 km from their manifolds.

The water depth varies between 300 m and 340 m and all equipment will be installed and retrieved without the use of divers.

34

R. WINDSOR AND P. A. NILSEN

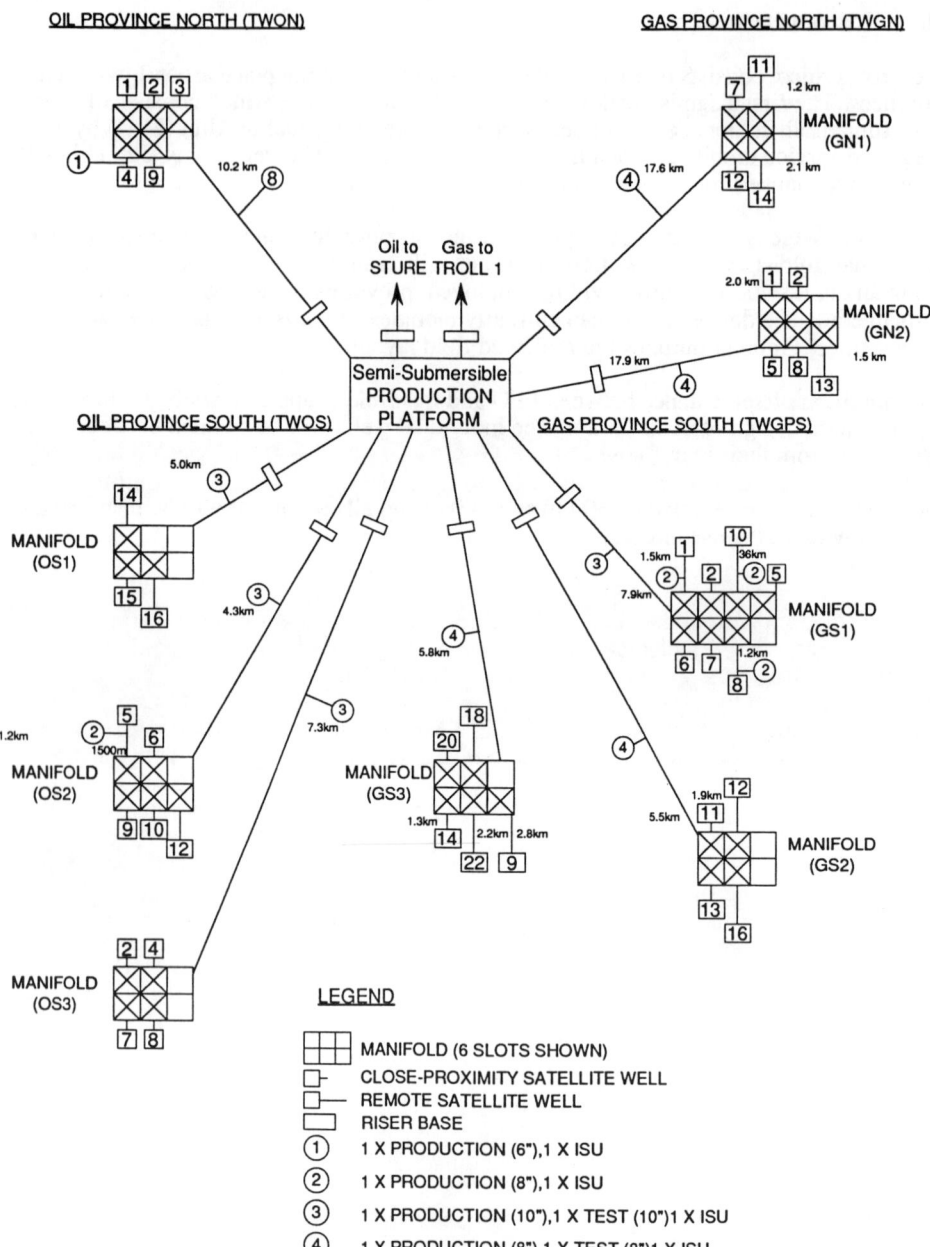

**Figure 1 - Troll Olje Development Plan**

## 2.0    System Definition

The Production Control System is an electrohydraulic multiplexed system which is designed for control of up to 24 wells in the oil province, expandable for connection of up to 32 additional wells in the gas province. The system consists of topsides equipment, located on the Floating Production Unit (FPU), which includes electrical power supplies and communications modems for supply and control of the subsea equipment. This is integrated with the FPU DISCOS system and is operated from the Central Control Room (CCR). The subsea equipment consists of electrical distribution jumpers/connectors, control pod and mounting bases located on the Xmas-trees, and transducers also mounted on the Xmas-tree for monitoring process pressures and temperatures. In addition, equipment is provided for installation and recovery of the Control Pod from the Xmas-tree and for testing purposes.

## 3.0    Surface Equipment

The surface equipment interfaces with a topside Subsea Control Unit (SCU) (supplied by others) via a series of communications I/O units. Electrical and electronic equipment is dual-redundant, each channel consisting of a power supply, three modems, three filter/power mixing units and a cross-connection cabinet, (4 units are used for manifold GSI only).

For each subsea manifold there is a dedicated Subsea Power and Communication Unit (SPCU), which is a 19" rack containing modems and power conditioning equipment for the subsea control pods. The electrical system is based upon the signal data being superimposed upon the power transmission lines in order to minimise cabling, particularly in the main umbilicals. The data encoding is accomplished through a technique known as Differential Quadrature Phase Shift Key or DQPSK, with the protocol based on the IEC TC 57 format. The combination of the DQPSK and protocol lead to a high integrity communications link with extremely low bit error rates.

Each SPCU is inherently dual redundant with two independent communications and power channels being supplied to each subsea control pod. However, a multidrop configuration means that each channel comprises a total of three modems and three filter/power conditioning units.

## 4.0    Subsea Equipment

### 4.1    Control Pod (Figure 2)

Each satellite tree is populated with a Control Pod (CP) (mounted on a pod mounting base) which is of a standard design, and fully interchangeable across the field. The CP is cylindrical with all connections (electrical and hydraulic) passing through the base.

Electronics are dual-redundant with two physically separate Subsea Electronic Modules, each comprising a 1-atmosphere pressure-isolating housing. Hydraulic supplies are single channel, but provision is made for a spare supply to enter the control pod via a spare hose in the umbilical.

The incoming low pressure supply is filtered and distributed to a bank of ten manifold-mounted Directional Control Valves (DCV's). The outputs from these DCV's control either tree valve actuators or pass back along the In-field ISU to control manifold-mounted pigging valves.

The incoming high-pressure supply is also filtered and distributed to a manifold populated with two DCV's which control Downhole Safety Valves.

Dual-redundant pressure sensors are incorporated for supply and return line pressure measurement and for inferred measurement of valve position. Dual redundant flowmeters are also incorporated for supply and return flow measurement.

The hydraulic couplings in the base are female, of the National Coupler type. Electrical connectors are two-contact female, conductive, controlled-environment type from Lockheed. Two Electrical connectors supply power and communications to/from the Control Pod and 8 gather signals from sensors mounted on the xmas tree. The CP design incorporates provision for flushing the electrical connectors in the base, prior to final make-up, using the Control Pod Running Tool (CPRT).

The CP is located via a pair of alignment pins on its base and secured by means of a single, central latch (Retlock). This is operated at the top of the pod, via a gearbox on the Control Pod Running Tool.

4.2      Control Pod Mounting Base (CPMB)

The CPMB becomes fully integrated into the Xmas-tree structure prior to deployment. It consists of a top plate which houses all mating hydraulic couplers and electrical connectors for the Control Pod, and the "anchor" to lock the control pod onto the CPMB via the "Retlock" latch.

Hydraulic connections from the CP pass via National Couplers into the CP mounting base and are then hard-piped either to the tree valve actuators or the manifold valve actuator (via the infield ISU).

Electrical connections are similarly routed, via oil-filled harnesses, to either the tree transducers or to the ROV connection point on the Infield ISU pull-in head. The two main electrical jumpers for comms-on-power are capable of remote connection and disconnection at both ends to facilitate replacement but this requirement does not extend to the transducer jumpers.

In addition to providing the means for locking the control pod and giving hydraulic and electrical connections, the CPMB is the main interface for the Control Pod Running Tool (CPRT). It has two guide posts for primary guidance of the CPRT onto the Xmas-tree; It provides the "buffers" on which the soft landing system of the CPRT react; and it provides the means by which the CPRT can be latched and delatched to/from the Xmas-tree.

4.3      Control Pod Running Tool (CPRT)

The Troll project has a dedicated running tool for CP installation and retrieval. The Running Tool and all operations associated with it are compatible with ROV intervention.

It is self-contained, not requiring any services via a surface-to-subsea umbilical and does not require fluid or electrical power to be supplied from the ROV.

The CPRT performs a number of functions:

* It has two guide funnels which give primary guidance of the CPRT or CPRT/CP onto the Xmas-tree via dedicated guide posts on the CPMB.

* It has a soft landing system designed to absorb impact loads during CP installation, thereby protecting the Control Pod.

* It has an ROV operated gearbox which activates the Retlock Latch in the CP to lock the CP to the CPMB on the Xmas-tree.

* It is capable of being locked onto the Xmas-tree (via the CPMB) enabling the CP
to be lowered hydraulically, and eliminating any effects from wave motion.

* It has the facility to flush the CP skirt with di-electric fluid, creating a benign environment just prior to make up of the electrical connectors.

4.4    Tree Sensors

Combined pressure and temperature sensors are provided for measurement of production and annulus pressure, and downstream choke pressure. All sensors are dual-redundant and employ a frequency-modulation technique for high accuracy.

**5.0    Maintenance and Test Equipment**

A suite of deliverable maintenance and test equipment has been provided to enable items of control system hardware to be functionally tested off line. This suite of equipment includes:

* Control Pod Test Stand (with Xmas-tree sensor simulator)

* Electrical Test Unit

* Umbilical Cable Simulator

* Dummy Control Pod (incorporating manual control functions for tree valve actuators)

* Protocol Test Unit and Maintenance Unit

* Test Equipment Container

* Test and Flushing HPU

* Communications Test Unit (Company-supplied)

* Tree Sensor Test Unit

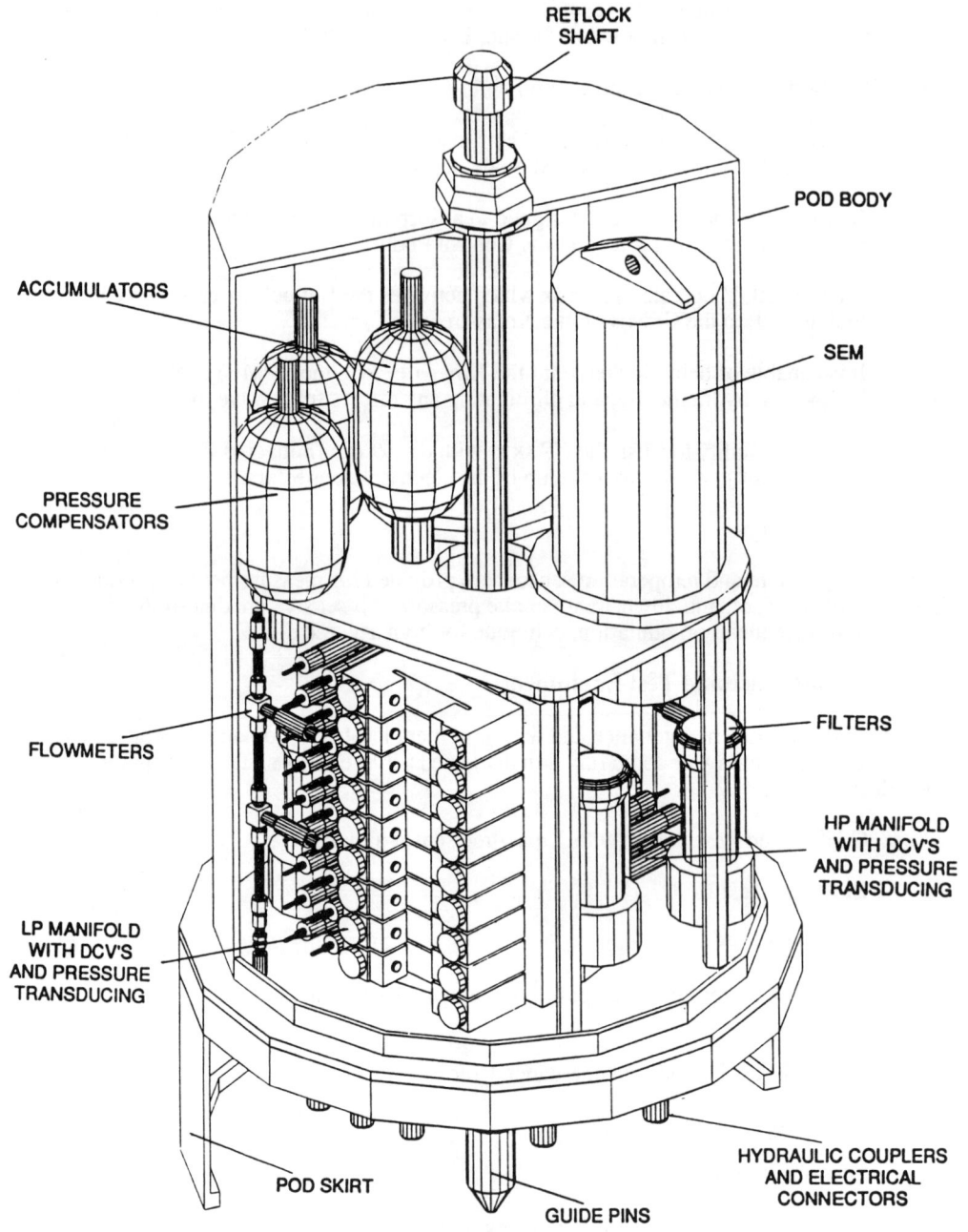

**FIGURE 2 – CONTROL POD**

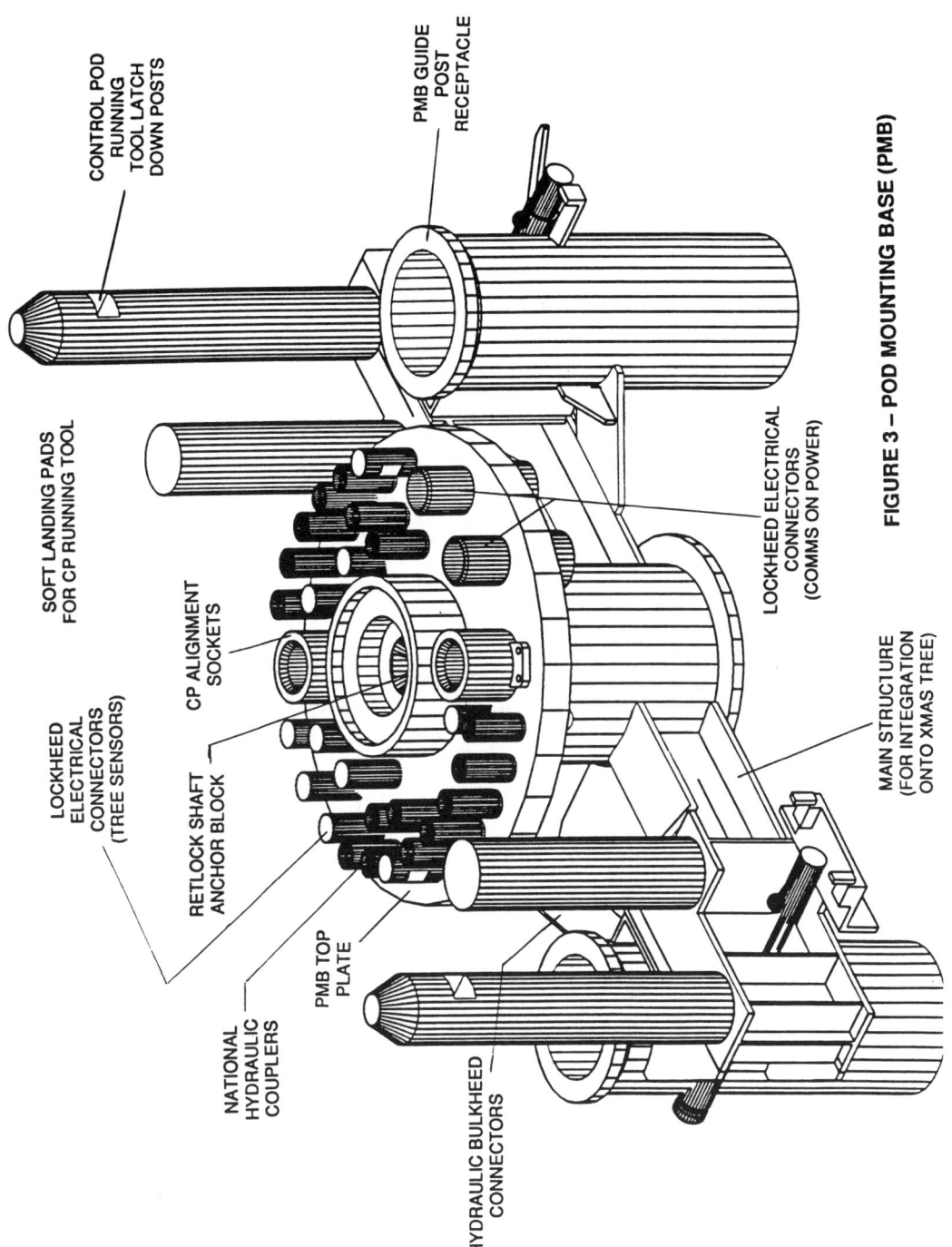

CONTROL POD RUNNING TOOL LATCH DOWN POSTS

PMB GUIDE POST RECEPTACLE

SOFT LANDING PADS FOR CP RUNNING TOOL

CP ALIGNMENT SOCKETS

LOCKHEED ELECTRICAL CONNECTORS (TREE SENSORS)

RETLOCK SHAFT ANCHOR BLOCK

PMB TOP PLATE

NATIONAL HYDRAULIC COUPLERS

HYDRAULIC BULKHEED CONNECTORS

LOCKHEED ELECTRICAL CONNECTORS (COMMS ON POWER)

MAIN STRUCTURE (FOR INTEGRATION ONTO XMAS TREE)

FIGURE 3 – POD MOUNTING BASE (PMB)

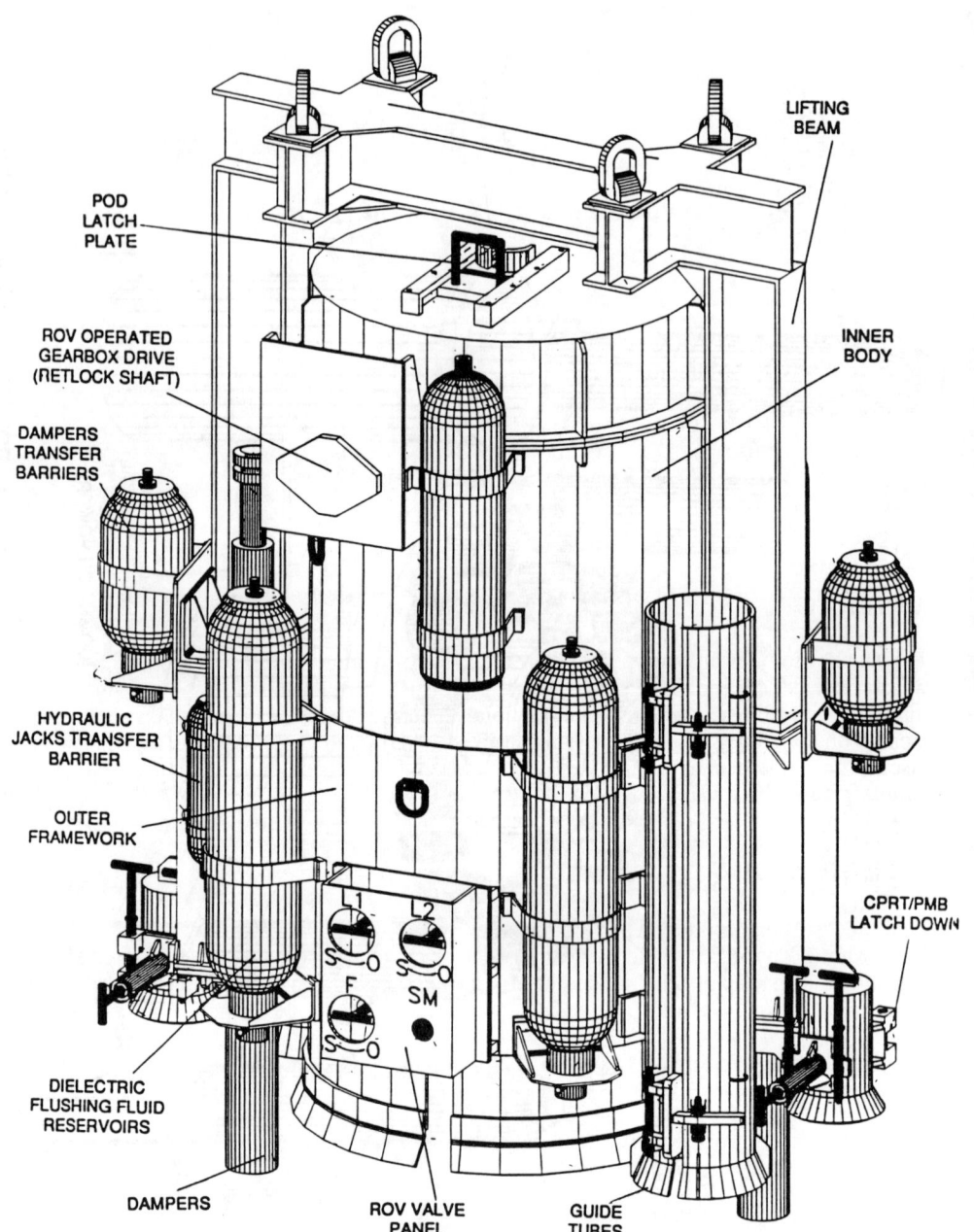

LIFTING BEAM

POD LATCH PLATE

ROV OPERATED GEARBOX DRIVE (RETLOCK SHAFT)

INNER BODY

DAMPERS TRANSFER BARRIERS

HYDRAULIC JACKS TRANSFER BARRIER

OUTER FRAMEWORK

CPRT/PMB LATCH DOWN

DIELECTRIC FLUSHING FLUID RESERVOIRS

DAMPERS

ROV VALVE PANEL

GUIDE TUBES

**FIGURE 4 – CONTROL POD RUNNING TOOL (CPRT)**

# EXXON'S ZINC DEVELOPMENT - OVERVIEW OF THE SUBSEA CONTROL SYSTEM

G R CLARK
GEC-Marconi
Oil & Gas
2 High Street
Nailsea
Bristol BS19 1BS
UK

M A STAIR
South Eastern Division
Exxon Co. USA
PO Box 61707
New Orleans
La 70161-1707
USA

## ABSTRACT

This paper provides an overview of the Zinc Electro-hydraulic Multiplex Control System, which provides for monitoring and control of 10 subsea wells and the subsea manifold functions as well as control and safety monitoring hardware for all Zinc process equipment located on the Alabaster platform. A description of the system architecture, together with a discussion of the design requirements, is provided, with an emphasis on the subsea template based hardware.

## INTRODUCTION

Zinc[1] is a subsea production system, installed in 1,460 ft of water, for development of the Mississippi Canyon 354 Field in the Gulf of Mexico . The subsea template is linked to Exxon's Alabaster platform, located in Mississippi Canyon Block 397 approximately 6 miles west of Zinc's location, via separate electrical and hydraulic control umbilicals.

The Zinc Subsea Production Control System is characterised by a number of significant design features:

- All subsea control system components located on a self-contained control skid which is integrated into the production manifold.

- Control and safety monitoring of all platform-based Zinc process equipment integrated with the subsea production control system.

- Guidewire installed subsea system components using a vessel of opportunity.

- Subsea intervention/maintenance by a work class 'ROV of opportunity'.

41

© 1994 *Society for Underwater Technology. Printed in the Netherlands.*

The control system architecture is implemented as an Electro-Hydraulic Multiplexed system with dual electric and hydraulic supplies consolidated into Subsea Control Pods (SCP's) for Xmas trees and manifold functions. Each interchangeable SCP, however, provides control and monitoring functions for either two trees, or one tree plus the Zinc subsea production manifold. All subsea hydraulic and chemical supplies are routed from the umbilical end terminations to the tree and manifold functions via fully welded hard piped tubing assemblies with metal-to-metal, self-sealing couplings being used at all disconnection points. Electrical connections between the incoming umbilical and distribution network to individual Control Pods are via ROV manipulator-deployed inductive connector assemblies. The entire template electrical distribution network with its supporting framework is recoverable using a simple Running Tool.

Equipment designed for the Alabaster platform follows industry standard practice; the operator's control console is dual redundant, hot standby with full colour graphic displays. In addition, significant integration of the control and safety monitoring of the associated Zinc process equipment has been achieved.

## SYSTEM OPERATING PHILOSOPHY

The Zinc Subsea Control System is an electro-hydraulic multiplexed system with Subsea Control Pods capable of controlling the tree valves on up to ten wells, plus the manifold pigging valves, and the chemical injection system. The control system also includes the capability to monitor/control Zinc-related Alabaster platform facilities via the Zinc Control Console (ZCC). The control system interfaces directly to the platform's existing safety/shutdown systems and automatically implements the various levels of both Emergency Shutdown and Process Safety Shutdown detected either by the platform's systems or the safety related topsides and operational subsea sensors which exist within the control system itself.

The Control System's subsea multiplexing is performed via the Subsea Control Pod, which is commanded by the Zinc Control Console. Each Control Pod is capable of operating two trees, or one tree and the manifold. The Control Pod switches hydraulic and chemical injection fluids by means of Solenoid-operated Directional Control Valves which are controlled by the Subsea Electronics Modules (SEM's) within the pod. The SEM's monitor input pressure transducers and output pressure switches on the hydraulic/chemical lines to provide feedback to the Control Console on the status of the system.

The subsea system is controlled via a dual-redundant serial data link with automatic selection by the ZCC of the active link. In the event that subsea electrical communications fail completely, then hydraulic pressure is automatically removed via the HPU, thereby causing all subsea valves to fail-safe-closed.

The subsea electrical system is powered by the platform-based Uninterruptible Power Supply (UPS) and Subsea Power Supply via independent dual-redundant power channels within the electrical umbilical. Each channel is normally powered and is consolidated into a single power supply within the individual Subsea Control Pods.

The hydraulic system operation is based upon surface-generated hydraulic supplies at 3,000 psi nominal for all tree valves, and 6,600 psi nominal for Surface-Controlled Sub-Surface Safety Valves (SCSSV's). These supplies are generated by the dedicated

Hydraulic Power Unit (HPU), which has redundant electric motor/pump systems, and duplicated HP and LP outlets.

At the subsea end, all hydraulic connections are by means of high-chromium 'super austenitic' steel tubing with welded fittings, and couplings to all retrievable modules being of a metal-to-metal, self-sealing type.

Some degree of distribution redundancy has been provided in the form of spare lines, and a retrievable Hydraulic Distribution Pod (HDP) with internally re-configurable piped connections.

The system is an open-loop hydraulic system, using 25% ethylene-glycol water-based control fluid. The hardware comprising the primary hydraulic control system has been designed to operate at a minimum fluid cleanliness level of NAS 1638 Class 10.

## Subsea System Configuration

On the Zinc template, the subsea control system hardware is located on a self-contained control skid at one end of the manifold. This configuration offers major advantages in terms of:

(a)   the potential for stand-alone system integration and test of the complete suite of control hardware (i.e. independent of manifold or template equipment).

(b)   the significant reduction in external interfaces with non-control system hardware.

(c)   independently installable and retrievable Electrical Distribution Network.

Key features of the subsea equipment are as follows:

•     Skid functionality has capacity to control up to 10 trees in addition to redundant facilities to control the associated manifold functions.

•     Separate electrical and hydraulic umbilical supplies which respectively interface to independently installable and retrievable Electrical Distribution Module (EDM) and Hydraulic Distribution Pod (HDP).

•     All hydraulic services hard-piped to the Hydraulic Umbilical Receiver Assembly, Xmas trees and manifold functions.

•     Inductive electrical connections, via a common design of ROV-connected jumper, between EDM and both the electrical umbilical and control pods.

•     Dual tree interchangeable Subsea Control Pods (SCP's) (i.e. able to control either two trees, or one tree plus a set of manifold functions), each installed using a dedicated dual guidewire ROV-operated Running Tool, on any one of 6 Pod Mounting Bases (PMB's).

•     Tall, minimum-diameter pod configuration (leading to compact skid design, able to be transported by road or rail).

- Ability to use non-dedicated work class 'ROV of opportunity' for all maintenance or intervention activities.

## EQUIPMENT SUMMARY - SURFACE

Located on the Alabaster platform (reference Figure 1) are:

- the Zinc Control Console (ZCC) which provides the control functions and operator interface for the subsea and surface system.

- the Hydraulic Power Unit (HPU) which supplies the high and medium pressure water based control fluids.

- the Uninterruptible Power Supply System (UPS) and Subsea Power Supply System providing electrical power for the Control Console and remote subsea template hardware.

- the Surface I/O Interface System (SIIS) which functions as the interface between the Control Console and above mentioned HPU and UPS, together with the associated Zinc Chemical Injection Unit, Process and Safety equipment.

In addition to the above primary equipment there are:

- Control Fluid Transportation and Storage Tanks which ensure that the cleanliness of the control fluid is maintained during handling operations between onshore supplier and offshore installation into the HPU.

- Subsea Emergency Shutdown System (SESD-3) which provides, on command, remote shut-in via the control system for all subsea wells from any drilling vessel located at the Zinc template.

## Zinc Control Console

The Zinc Control Console (ZCC) is the centre point and main operator interface for both the sub-sea and surface control of the Zinc facilities. The ZCC's basic functionality is operator-driven, sequence-driven and alarm-driven, all features of which are totally user-configurable.

The ZCC hardware consists of twin (main and hot standby) real-time micro-computers, each with an associated colour graphics system and keyboard, a hard disc for program and data storage and a tape drive for long-term storage of logged data, backup and software maintenance. A printer is provided for event logging, reports and graphic outputs.

Operators issue control commands at either the graphics console or annunciation/input panels to operate equipment. The operator can either nominate individual equipment to operate directly, or start a control sequence to operate a group of equipment. A control sequence is a list of commands that orders operation of any equipment to any process state. It allows conditional tests (e.g. greater-than, less-than and if-in-alarm, etc.) to change the flow of control or wait until an event occurs. Sequences are user-entered, and may be

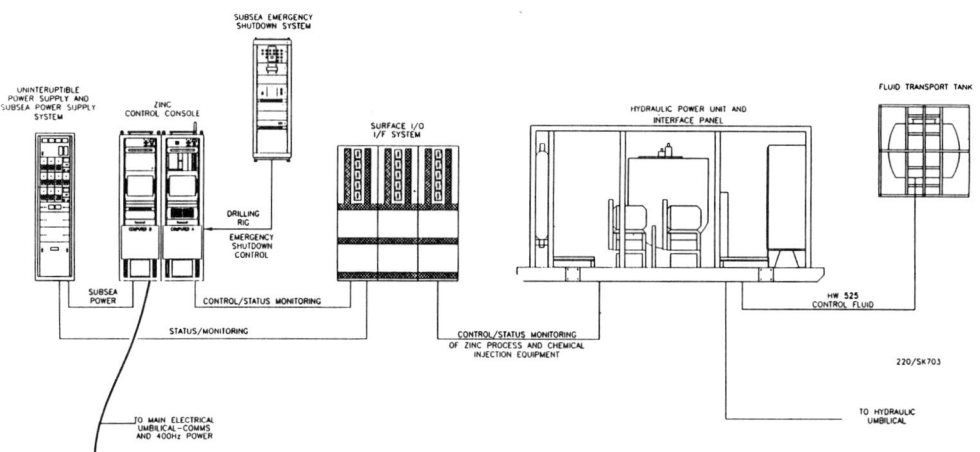

**FIGURE 1:  CONTROL SYSTEM TOPSIDES EQUIPMENT
— LOCATED ON ALABASTER PLATFORM**

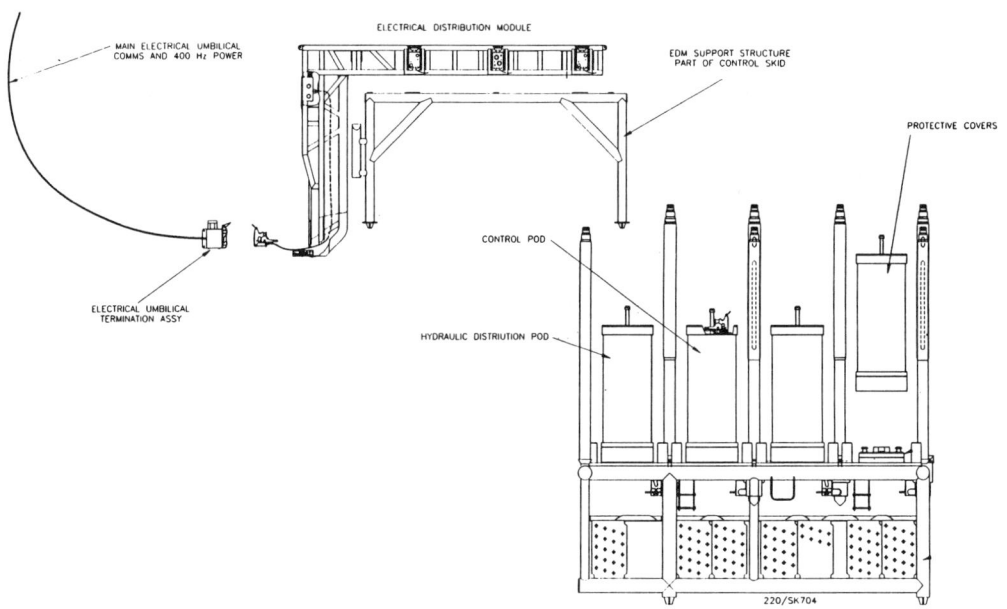

**FIGURE 2:  CONTROL SYSTEM SUBSEA EQUIPMENT
— LOCATED ON ZINC TEMPLATE**

entered or modified at any time during the operating life of the system.
Alarms can control a digital output, a secondary message, or start a control sequence.
Every input discrete has one alarm state; this can be either of its two states. Analogues have
six alarm states, three upper and three lower. Valves have two alarms, a failure-to-operate
alarm and a state alarm (either state); any of these alarms can operate control sequences.

## Sequences and Interlocks

The autonomous operation of the system is controlled by sequences. These are lists of
commands to operate equipment with conditional steps contained within to effect the timing
and flow of control through the sequence. Sequences that are started by operators have no
special significance other than their obvious control function. Any sequence started by an
alarm takes on a new role, that is, as an interlock, thereby ensuring that under no
circumstances can the facility be put at risk by accidental operator intervention.

## Surface I/O Interface System

The Surface I/O Interface System (SIIS) is the data collection and first line control of the
surface facilities. The operator interface for the SIIS is limited to simple push buttons and
indicators on annunciator/input panels.

The SIIS consists of four Programmable Logic Controllers (PLC's) (in a non-redundant
configuration). One PLC is concerned with Safety monitoring and is known as the "Safety
PLC", one is concerned with control and interlocks on the platform based Zinc facilities
and is called the "Control PLC" and the remaining two run PID control loops for the
Control PLC and are called the "Slave PLCs".

The four PLC's are connected together and to the active ZCC computer via a local area
network, which allows the ZCC to access to all transducer data on the PLC's and to
"program" transducer parameters such as alarm limits and PID gains and set points.

## EQUIPMENT SUMMARY - SUBSEA

Located at the Zinc template (reference Figure 2) are the:

- Control Skid (integral with the Zinc manifold) upon which all of the retrievable Control
  System hardware is located.

- Retrievable Electrical Distribution Module (EDM) which interfaces the main Electrical
  Umbilical from the remote Alabaster platform to each of the Subsea Control Pods
  located on the Skid.

- Subsea Control Pods (SCP's), up to six maximum, each of which provides all control
  and monitoring functions for either two trees, or one tree plus the subsea production
  manifold functions.

- Hydraulic Distribution Pod (HDP) which takes the incoming hydraulic power and
  Chemical Injection (CI) lines from the hydraulic umbilical and distributes them via
  control skid network tubing to the individual SCP's.

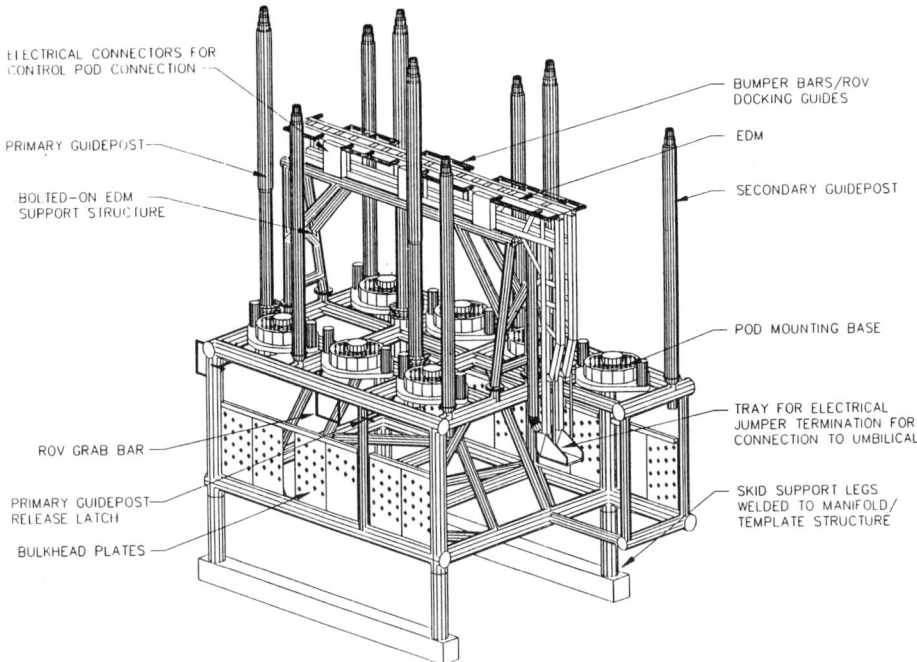

**FIGURE 3:  CONTROL SKID ASSEMBLY (EXCLUDING PODS)**

**FIGURE 4:  CONTROL SKID ASSEMBLY
— AS INSTALLED ON ZINC TEMPLATE**

In addition to the above, there are:

- Protective Covers (PC's) which are installed as required in each of the unused control pod locations.

- Electrical Umbilical Termination Assembly (EUTA) which is integral with the main electrical umbilical and provides the interface connection to the EDM.

## Control System Skid

The Control Skid itself (reference Figures 3 & 4) consists of a structural framework; this supports the Control Pod Mounting Bases (which interface with replaceable Control Pods) and the hydraulic distribution network (tubing). The hydraulic tubing passes from the Pod Mounting Bases (PMB's) to bulkhead plates at the perimeter of the skid, these plates providing the interface between skid and manifold tubing. The skid also incorporates docking and landing features to facilitate installation and maintenance operations on skid-mounted equipment e.g. ROV-removable guide posts, Electrical Distribution Module (EDM), Subsea Control Pods (SCP's), Hydraulic Distribution Pod (HDP), and Protective Covers.

Over 200 separate tubing connections interconnect the hydraulic distribution networks of the skid to the manifold. For simplicity and reliability, these are fabricated as permanent, all-welded assemblies. This approach means the Control Skid can only be retrieved with the manifold. As a result, individual components resident on the Control Skid that may be subject to failure, are all designed to be retrieved for maintenance.

Overall size of the skid is constrained to permit land transportation (road or rail) and to minimise weight and footprint area on the manifold/template. This resulted in a compact tubing layout of approximately 3,000 feet of hydraulic tubing in a skid which measures 23 ft L x 13 ft 6 in W x 7 ft H (excluding guideposts and EDM support structure), with tubing fabrication following a strict installation sequence developed using 3D CAD techniques.

Figure 3 shows the fully-assembled Control Skid without the Control Pods, which have been left out for clarity. Key design features of the Control Skid are:

- Simple and robust tubular framework to minimise weight and maximise rigidity.

- Simple weld points for attachment to manifold/template framework.

- Integral guidance and support provisions for all retrievable modules.

- Inner Guideposts, equipped with ROV-operable releases, and Outer Guideposts, with shear-out provisions, to facilitate recovery and replacement as required during maintenance activities.

- Super Austenitic hydraulic tubing (for long-term corrosion resistance) connecting Pod Mounting Bases to manifold control fluid distribution system through bulkhead connection plates.

- Integral cathodic protection.

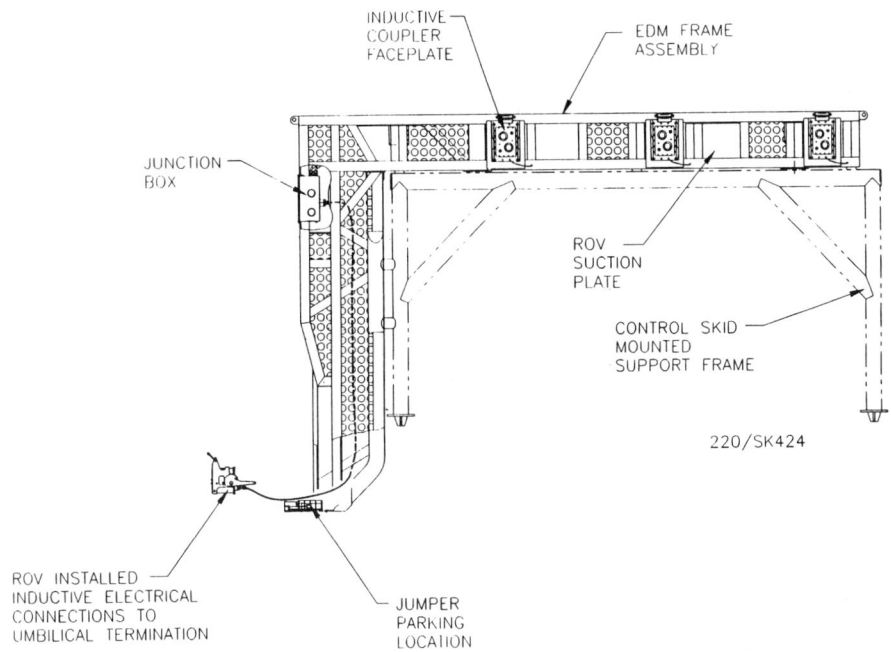

**FIGURE 5:   ELECTRICAL DISTRIBUTION MODULE (EDM) FEATURES**

**FIGURE 6:   EDM INSTALLATION / RETRIEVAL**

Design of the skid allows for maintenance of the pods, EDM and guideposts by complete removal and replacement , and also allows for routine inspection by ROV's.

**Electrical Distribution Module (EDM)** (reference figure 5)

The EDM is designed as a lightweight framework, supported on the Control Skid and spanning the distance between the Electrical Umbilical Termination and Control Pods. It provides a means of conveying electrical power and signals from the electrical umbilical to the pods.

The form of the EDM comprises an L-shaped, rectangular cross section open framework, with a vertical downstand of similar cross section at one end. Mounted in each vertical side of the horizontal section are sets of inductive electrical connectors, to which the ROV attaches jumpers from the skid-mounted Control Pods. Oil-filled flexible cable conduits from each connector run along inside the horizontal framework of the EDM and drop down the vertical downstand to a junction box. From this box, a jumper cable runs to an inductive Electrical Jumper Termination (EJT) parked at the bottom of the vertical downstand of the EDM framework. The EJT, which shares a common design with the control pod connections, is connected to the main Electrical Umbilical Termination by ROV.

Inductive connections were selected because of their high reliability and tolerance to mechanical misalignment. Also, the inclusion of a live-disconnect feature is a major benefit, in that it allows a pod to be disconnected without in any way affecting operation of the rest of the system and provides automatic fault isolation downstream of the EDM. Dual-channel independent power and communications systems are provided.

The EDM is normally located and latched onto a support framework on the control skid structure. In the event of significant physical damage to the EDM and/or its support structure, it is possible to retrieve it (or them) to the surface and replace them by a suitably-modified EDM which could interface with secondary lockdown points provided on the Control Skid. The EDM also has provision to accommodate transponders to facilitate ROV operations.

**Subsea Control Pod (SCP)** (reference Figs 7 & 8)

Key design parameters of the SCP, in addition to those previously described, are as follows:

• Thirty year design life target.

• No flexible hydraulic jumper connections; all hydraulics base-connected suitable for diverless operation.

• Dual-channel electrical power supplies and communications channels, consolidated within SCP, in individual electronics modules.

• Dual-channel hydraulic supplies (consolidated within SCP).

FIGURE 8:  SUBSEA CONTROL POD (SCP)

FIGURE 7:  SUBSEA CONTROL POD FEATURES

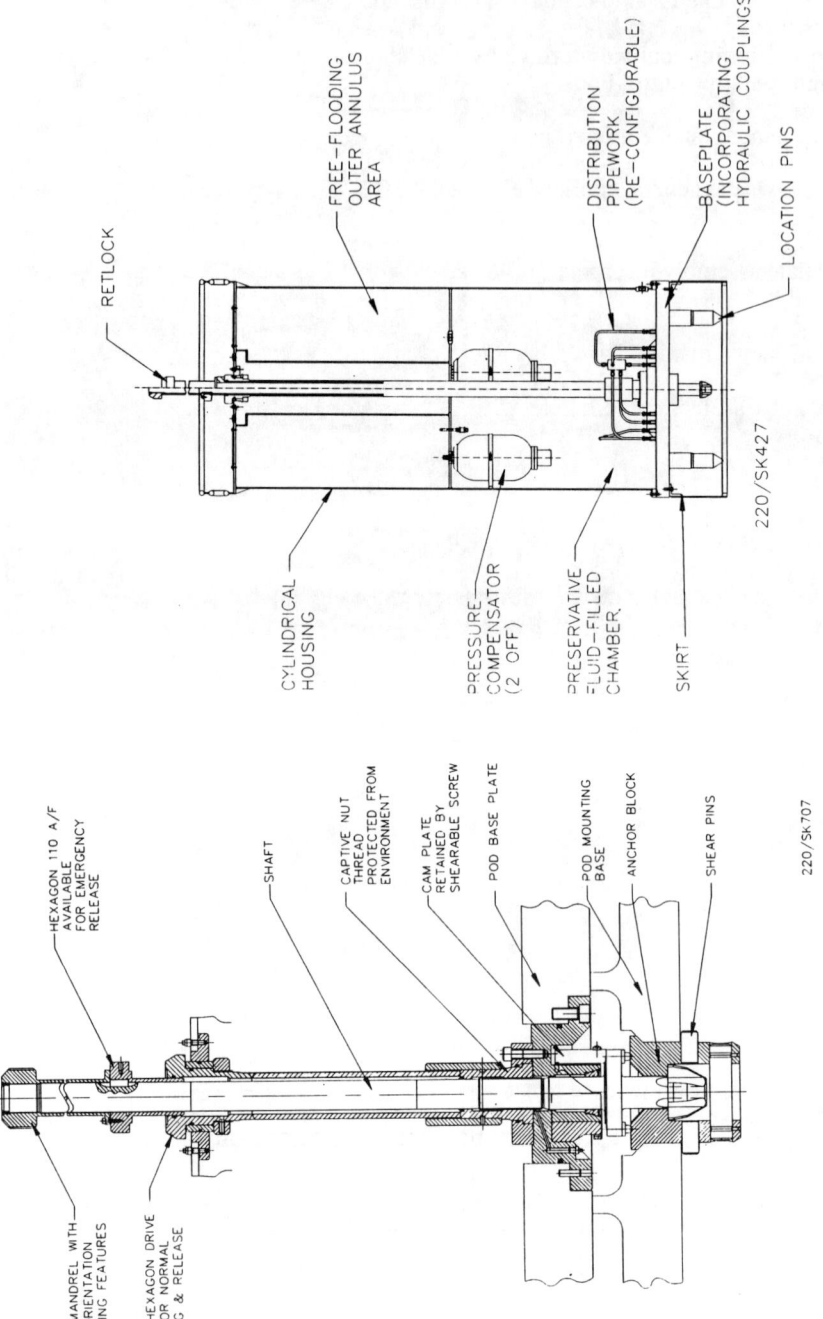

220/SK427

FIGURE 10: HYDRAULIC DISTRIBUTION POD (HDP)

220/Sk.707

FIGURE 9: RETLOCK LATCH ASSEMBLY

- Provision for chemical injection control functions.

- Provision for internal monitoring of wellhead pressures (i.e. wellhead pressures brought within Control Pod).

- Single, central latch configuration.

Figure 7 shows the general configuration of the SCP. Significant features are identified as follows:

- Baseplate assembly incorporating 46 stab-type metal to metal, self-sealing hydraulic couplings.

- Central "core" manifold assembly (dielectric fluid-filled).

- Twin Subsea Electronic Modules (SEM's) - one for each set of tree (or manifold) functions.

- Internal support stanchions to which the methanol injection control valves are attached.

- Central 'Retlock' pod latchdown mechanism.

- Top-mounted captive inductive electrical jumper assembly for ROV connection to EDM.

- All remote intervention operations carried out at top of pod.

## The 'Retlock'[2] Latching System

The 'Retlock' concept was initially conceived and developed as a means of remotely installing and retrieving subsea hardware in such a way that no part of the operating mechanism would remain subsea.

For this application, the Retlock has been used to secure the interface between the SCP (or HDP) and the Pod Mounting Base on the Control Skid. The pod is run using a dedicated Running Tool which, upon landing, provides the torque to make up the hydraulic connectors, and then apply a pre-load to resist the effects of vibration, uneven loads across the interface, and hydraulic separation forces.

The Retlock (reference Figure 9) is unique in that it embodies the tried and trusted bolt and nut fastening method with additional features to enhance performance and reliability subsea. These features are:

- It is not possible to tighten the Retlock without first achieving the proper alignment of mating components. Visual alignment markers are provided for ROV-inspection to confirm correct Retlock engagement.

- All threads and active components are retrievable (with SCP, HDP or Protective Cover).

- A simple, robust profile is all that remains in the non-retrievable portion.

- Retlock is capable of creating large push-apart (hub separation) force to break hydraulic lock, or overcome component seizure.

- Retlock shaft can withstand several times pod self-weight, so that base connections are protected during landing. Coupler make-up and disconnection is slow and controlled.

- In the unlikely event of thread or bearing seizure (freezing), it can be overcome by application of over-torque directly onto the Retlock shaft to ensure secondary release.

A special feature of the Zinc Retlock assembly is a tertiary release mechanism in which, should both the primary and secondary releases fail, then a Control Pod can be retrieved using the Emergency Release Running Tool (ERRT). This applies a tensile pull to the Retlock's lifting mandrel and shears the anchor block out of the Pod Mounting Base. After replacing its shear pins, the anchor block can be re-installed into the Pod Mounting Base, during Control Pod replacement.

## Hydraulic Distribution Pod (HDP)

The function of the HDP (reference Figure 10) is to accept incoming hydraulic services from the main surface-to-subsea hydraulic umbilical, and distribute them to supply headers on the Control Skid. It too may be installed/retrieved using the same Running Tool as for the Control Pod.

The salient feature of the HDP is that it provides flexibility in the configuration of the hydraulic distribution system in that the pod, once retrieved to the surface, may be internally re-configured to overcome either a failure in the control skid-mounted piping by means of utilisation of spare lines in both skid and incoming hydraulic umbilical, or to implement revised distribution requirements. The HDP is packaged in such a way that the external dimensions and interfaces allow it to be treated as if it were a Subsea Control Pod. Thus, common intervention procedures and tooling are employed.

## Protective Cover (PC)

For skid positions unoccupied by an SCP, Protective Covers (PC's) are provided. This unit has the same mechanical outline and interfaces as the Subsea Control Pod and serves the dual function of protecting the exposed Pod Mounting Base hydraulic couplers and terminating any couplers that may be pressurised during system operation.

## TEST AND MAINTENANCE EQUIPMENT

A suite of equipment, to assist in the maintenance of the control system, includes the Pod Running and Emergency Retrieval Tools, EDM Running tool, together with a Test Stand and Test Hydraulic Power Unit which, with a portable Command/Test Unit (CTU), provides the facility for onshore or rig-based maintenance of either the SCP, HDP, or PC.

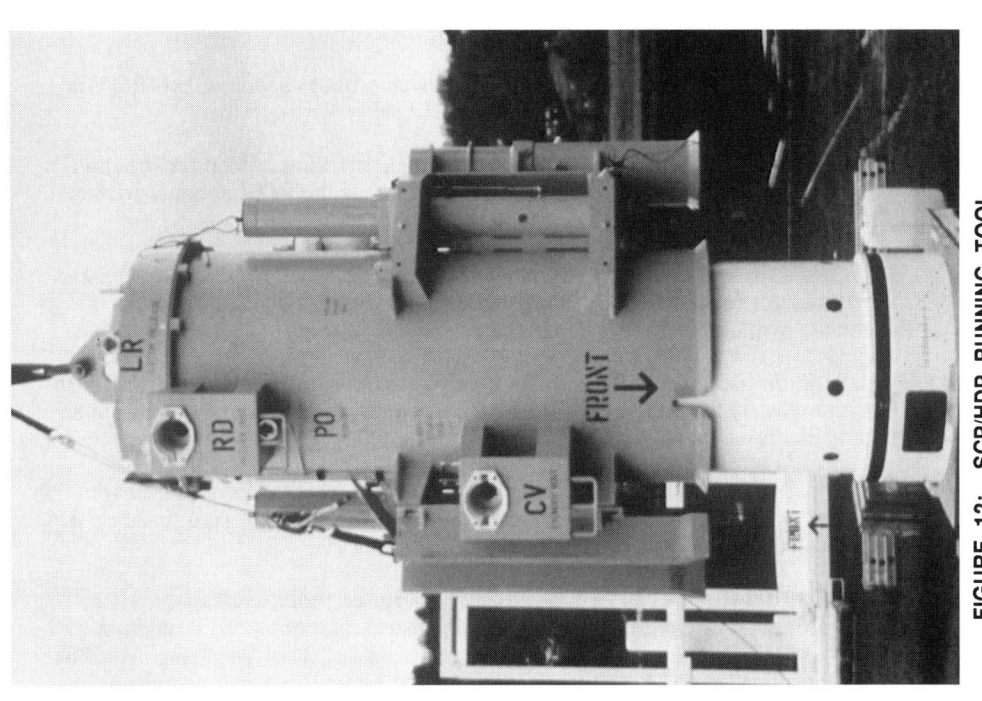

**FIGURE 12: SCP/HDP RUNNING TOOL**

**FIGURE 11: SCP/HDP RUNNING TOOL FEATURES**

## Control System Maintenance - Subsea Equipment

The maintenance of the control system's subsea hardware utilises a work class 'ROV of opportunity' for:

- inspection operations - where ROV access, equipment marking , metal finishes and paint colourations were considered during design to assist the ROV operator in these tasks.

- connection and disconnection operations - which were designed to be executed using ROV manipulator functions, ROV hydraulic supplies through hot stabs and commercially available ROV torque interface.

For the installation/retrieval operations of the various control system modules (e.g. EDM, SCP, HDP, and covers) the work class 'ROV of opportunity' is supplemented by stand-alone Running Tools. Three different Running Tools are provided as follows:

- EDM Running Tool (Figure 6) - a simple beam structure which is used in the event that the EDM or/and its support structure incur either a serious failure or significant damage sufficient to prevent the control system from functioning.

- Pod Running Tool (Figures 11 & 12) - an ROV energised tool used for installation/retrieval of Control Pods, Hydraulic Distribution Pods and Protective Covers. The design intention to create a 'passive' Running Tool, requiring no surface-to-subsea umbilical, no on-board electronics, and no hydraulic motors or pumps, was successfully achieved with the additional advantage that the tool does not necessarily require a surface deployment vessel equipped with heave compensation.

- Emergency Release Running Tool - an ROV-energised tool used for the retrieval of either pods or covers, should both their primary and secondary pod latch release mechanisms fail.

The utilisation of these tools, together with intervention operations of the ROV, are now described, to illustrate the relative simplicity of maintenance of the subsea control equipment.

## Intervention Systems and Hardware

A simple set of arrangements has been configured for the ROV to dock and perform various operations on the control skid equipment. All docking arrangements have been designed as systems without permanent latching arrangements. Docking arrangements for principal ROV tasks are as follows (reference Figure 3):

## ROV-Operated Electrical Connections Between EDM and Either Control Pods or Electrical Umbilical Termination Assembly (EUTA)

Bumper bar/ROV docking guides are provided along the perimeter of the EDM framework and EUTA at their top surfaces. These allow the ROV to rest on the top surfaces of either the EDM or the EUTA and thrust continuously downwards to hold position. The ROV can be simply lifted off, should it suffer any power failure during intervention operations.

For SCP to EDM connection, the ROV, in the above docked mode on top of the EDM, removes the protection plate from the EDM inductive connection faceplate. It then reaches out with its manipulator, lifts the inductive Electrical Jumper Termination (EJT) off its stowage position on the top of the SCP, and brings it in towards the EDM. The EJT is then inserted into the EDM connection point, and latched. Similarly, when docked on top of the EUTA, the ROV performs an identical sequence of operations to connect the EDM to the EUTA.

## SCP/HDP Installation/Retrieval

As part of the design philosophy, the efficient use of tooling has been of paramount importance. The aim has been to produce a cost-effective solution by maximising tooling interchangeability and minimising the complexity of the tool design.

This philosophy has led to a multi-purpose Running Tool concept, capable of installing and retrieving Subsea Control Pods, the Hydraulic Distribution Pod and also the Protective Covers.

The tool is designed to run from the surface using a guidewire system connected to the Control Skid guideposts. The Running Tool (reference Figure 11) runs on the guidewires and orientates on the guideposts. The Running Tool/SCP combination is brought to rest by a soft landing system comprising dual dampers.

ROV intervention is then required to operate a cylinder Vent Control valve which initiates controlled descent of the Running Tool/SCP combination for the final engagement to the Pod Mounting Base. The 'Retlock' drive system, which forms an integral part of the Running Tool, is operated by ROV until a positive latchdown to the Mounting Base is achieved. The prime motive force for the Retlock drive is supplied as a torque input from the ROV torque tool.

Once fully engaged, the SCP/HDP (or Pod Mounting Base Cover) is unlatched from the tool by the ROV. This system uses a spring-loaded latch system which latches under the lifting mandrel on the SCP/HDP (or cover) to the tool; the latch being operated by ROV-supplied hydraulics, with a mechanical override.

The Running Tool also incorporates a hydraulic "hot-stab" system for the ROV to supply pressure to hydraulic cylinders for both testing the fastness of the Retlock latchdown, and also for jacking the SCP/HDP off the anchor base during retrieval operations. Visual indication of the position of the Retlock locking head with respect to its anchor block, and also the position of the SCP/HDP latch system, is available for the observing ROV.

## SUMMARY AND CONCLUSIONS

Key influences in the development of the Zinc control system's design have been:

- providing a control system based on field-proven technology, techniques and procedures.

- minimising subsea intervention/maintenance through fault tolerant design and, when required, that such ROV/workover operations are neither complex nor require sophisticated techniques.

- designing all subsea electrical components to be retrievable for maintenance independent of retrieving the subsea manifold.

The significant achievements, in these respects, have been:

- Integrated topsides process and subsea control from the Zinc Control Console.

- A centralised and integrated subsea hardware design, capable of being assembled and tested independently from the template/manifold.

- A compact base-area Subsea Control Pod which combines high functionality with straightforward remote intervention and retrieval procedures.

- Separately installable and retrievable electrical and hydraulic distribution modules, allowing reconfiguration of interconnections to cope with fault conditions, or system enhancements.

- Simple, cost-effective, multi-purpose design of passive running tools for the installation and retrieval of SCP/HDP, Protective Covers and Electrical Distribution Module.

- Development of practical, efficient methods of deepwater intervention utilising general-purpose ROV's and workboats as the primary means of intervention.

- Elimination of positive-lock latching devices between the ROV/ROV tooling and subsea equipment (ROV/ROV tooling is always retrievable in the event of equipment malfunction without the need for complex override mechanisms).

- Integral and simple mechanical back-up features to provide at least one level of redundancy for long-term subsea connections, with minimum complexity and cost.

In conclusion, the Zinc Subsea Production Control System represents a technical implementation of a state-of-the-art system configuration but with enhanced reliability features which are justified by the depth of field application. The favourable outcome of the extensive land integration and test activity (during 1992) fully supports confidence that the control system will live up to its performance expectations. Close client/contractor collaboration throughout conceptual and detail design, fabrication and testing, and onshore integration and testing was necessary to achieve this result.

## NOMENCLATURE

| | |
|---|---|
| CAD | Computer Aided Design |
| CI | Chemical Injection |
| CTU | Command/Test Unit |
| EDM | Electrical Distribution Module |
| EJT, | Electrical Jumper Termination |
| ERRT | Emergency Release Running Tool |
| EUTA | Electrical Umbilical Termination Assembly |
| HDP | Hydraulic Distribution Pod |
| HP | High Pressure |
| HPU | Hydraulic Power Unit |
| LP | Low Pressure |
| PC | Protective Cover |
| PID | Proportional Integral and Derivative |
| PLC | Programmable Logic Controller |
| PMB | Pod Mounting Base |
| ROV | Remotely Operated Vehicle |
| SCP | Subsea Control Pod |
| SCSSV | Surface-Controlled Sub-Surface Safety Valve |
| SEM | Subsea Electronics Module |
| SESD-3 | Subsea Emergency Shutdown System |
| SIIS | Surface I/O Interface System |
| UPS | Uninterruptible Power Supply |
| ZCC | Zinc Control Console |

## ACKNOWLEDGEMENT

Copyright 1993, Offshore Technology Conference

This paper is based on the following paper, which was presented at the 25th Annual OTC in Houston, Texas, U.S.A., 3-6 May 1993:

OTC 7285    Zinc Project:   Overview of the Subsea Control System.

## REFERENCES

1. Bednar J. M. - "The Zinc Subsea Production System - An Overview" - OTC Paper No. 7283 presented at the 25th Annual OTC in Houston, Texas, May 3-6, 1993.

2. Pejaver D. R., Jones J. W., and White J.: "Diverless Maintained Clusters 'DMaC' Subsea Production System" - OTC Paper No. 6720 presented at the 23rd Annual OTC in Houston, Texas, May 6-9, 1991.

# Session 2
# Sensor and Component Development

# A ROBUST SOURCE FOR UNDERSEA FIBRE OPTIC SENSOR SYSTEMS

Dr. E. Lewis(+), Mr. W.R. Dutch(+), Dr. P.J. Scully(+) ,
Mr. B.J. Deboux(+) and Mr. K.J Green (‡)

+ Optical Fibre Sensors Research,     ‡ Design Engineer,
  Liverpool John Moores University,  Industrial Control Services PLC,
  Byrom St.,                          Maldon,
  Liverpool L3 3AF.                    Essex CM9 7LA.
  United Kingdom.

## ABSTRACT

A low cost and robust temperature controlled optical source for fibre optic measurement systems has been developed. The temperature control has been achieved by means of an active heat sink thermally coupled via a copper block housing. An embedded 80188 system with the appropriate Input / Output boards has enabled a flexible and low cost controlled source to be realised. The hardware is mounted in a 3 slot STE BUS miniature rack which includes the PSU. The software was developed on a 286 host and downloaded to the 80188 card using Sourceview [1] software. The use of Borland C++ as a programming language has enabled relatively simple PID Control Algorithms to be developed, though more sophisticated ones may be incorporated in the future.

The control system accuracy has been demonstrated as better than 0.05 $^{\circ}$C with a high degree of long term stability with tests of many hours duration having been conducted. The instrument is configured to enable the wavelength and/or the intensity of the source to be modulated . This provides the added advantage of being suitable for a wide range of  sensor applications.

It is considered that this high performance, low cost and robust instrument is ideal for applications such as optical fibre sensors which could be used in undersea environments. Additionally the system is easily adapted for use as stabilisation of temperature sensitive detectors e.g. PIN diodes and APDs. A range of sources and detectors can be easily incorporated into this system with fibre coupling being achieved by connectors to standard device housings or direct fibre pigtailing.

*Volume 32: Subsea Control and Data Acquisition, 63–80.*
© 1994 *Society for Underwater Technology. Printed in the Netherlands.*

## INTRODUCTION

The instrument developed as a result of the work reported in this paper is suitable for a wide range of fibre optic sensor measurement systems including those which can be used for subsea applications. The instrument is capable of housing LEDs (Light Emitting Diodes) Laser Diodes and/or Photodetectors in a robust and transportable unit. This makes it particularly suitable for subsea applications which involve remote measurement e.g. pressure, temperature, displacement by fibre optics. The general use of fibre optics for subsea systems is wide ranging and well established [2,3,4]. Advantages offered by fibre optics for subsea monitoring and measurement include immunity from effects of a harsh operating environment and safe operation in potentially hazardous environments e.g. explosion risk. Much research has been directed towards the development of subsea fibre optic systems. However, one of the major practical limitations of such systems has been the long term performance of the light source and to a lesser extent the detector. All LEDs and Laser Diodes exhibit performance limiting characteristics e.g. drift of output power due to temperature and ageing (time) [5,6]. These factors also have an influence on photodetectors e.g. PIN photodiodes and APDs through variations of the dark current. The instrument developed in the work described can accommodate light sources and detectors simultaneously. However, the bulk of the experimental results were for a 780 nm (nominal centre wavelength) Laser Diode which represents the most severe experimental conditions for stabilisation.

Temperature stability was achieved by the use of an active heat sink (Peltier Heat Pump) which is driven from a conventional electronic amplifier power output stage. Control was established using PID algorithms resident on a target 8088 based embedded system. The development using embedded software rather than a full hardware implementation was intended to promote a system which was highly versatile and easily re configurable. This approach was necessary so that the full range of sources and detectors were available for testing. Furthermore this presented the possibility of incorporating other types of control algorithms e.g. PD, PI or more sophisticated ones including fuzzy logic operations [7].

Several tests have been performed on the system to evaluate its fundamental performance in terms of its ability to stabilise the device temperature, and hence in the Laser Diode case maintain constant output Light Intensity and wavelength distribution. Some of the results of those tests are reported here for the example of PID control on the 780 nm Laser Diode. It will be shown that the system achieves a high degree of accuracy coupled with good long term stability.

Versatility in terms of the type of optical device housed in this instrument is clearly significant if a wide range of sensing applications are to be addressed. However, it also important to ensure that the device can be operated in different modes e.g. Continuous Wave (CW) or Pulsed output. Furthermore, it is demonstrated that it is possible to accurately modulate the wavelength distribution of the Laser Diode source so that this to may be used as a means of interrogating fibre optic sensor systems.

The establishment of such a controllable source is significant in terms of its possible use for other more sophisticated sources e.g. Fibre Lasers [8,9,10]. The latter devices have been proven to offer enhanced light output characteristics (Intensity and wavelength) for certain types of fibre sensor systems. The relation of the present instrument to these is also discussed in this paper.

## CONFIGURATION OF THE DEVELOPMENT SYSTEM

The experimental configuration for the development of the stabilised source is shown in Fig. 1. It comprises four basic units, namely the optical source (780 nm Laser Diode), the host PC (an Arcom 286 based microprocessor development system), the unit housing the target system and the Power Supply Unit.

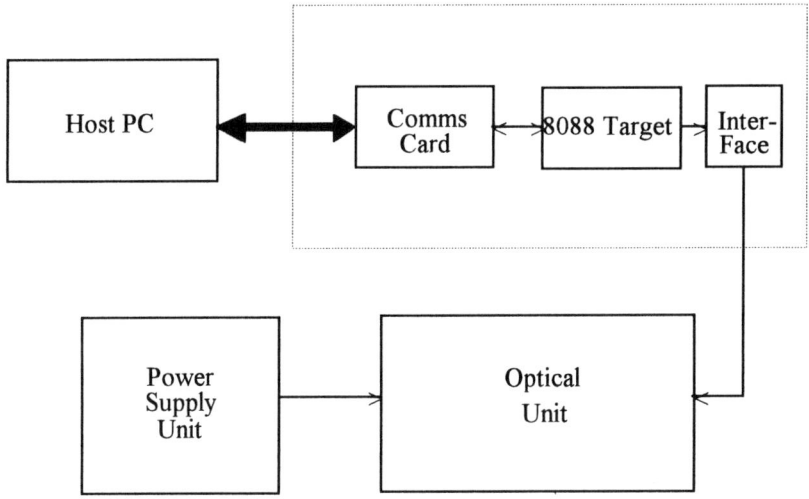

Figure 1. The Development System.

The basic units and their functions are described below :-

## The Optical Unit.

This consists of an actively cooled copper block which houses the Laser Diode Source (or other optical device). Cooling and heating of the copper block is achieved by means of a Peltier Heat Pump sandwiched between the copper block and an appropriately sized heat sink. The physical layout is shown in Fig. 2. The method of operation of the Peltier Heat Pump is relatively simple and involves the passage of current in different directions in order to achieve the required heating or cooling effect

. When current is passed through it in one direction it pumps heat from one side to the other (passage of heat from the copper block results in cooling) and when the current is in the opposite direction the heat is pumped in the other direction (passage of heat to the copper block results in heating). In this way it is possible to control the heat flow to and from the copper block (and thus control its temperature). This is achieved by controlling the direction and amount of current passed through the heat pump using a power amplifier.

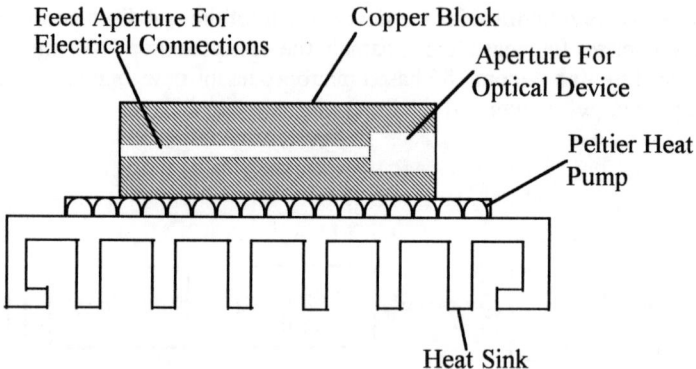

Feed Aperture For
Electrical Connections

Copper Block

Aperture For
Optical Device

Peltier Heat
Pump

Heat Sink

Figure 2. The Physical Layout Of The Optical Unit.

**The Host PC**

This used solely for development purposes and forms no part of the final version of the instrument. The hardware comprises an Arcom STE bus microprocessor development system which is modular and includes a 286 processor card (with on board disk controller, graphics driver etc.), a power supply unit and a communications card. This system is configured to imitate the 286 based IBM PC and with the right development software a PC (286 or higher level) could be used in the same way. The STE bus system was chosen because of its robustness (STE bus is an industry standard for computer back-planes and plug in card dimensions for control and instrumentation applications). The development software used was Borland C which was downloaded to the target system using the Arcom Sourceview [1] software. This approach allowed maximum flexibility in the initial software development making use of all of the debugging facilities of Borland C prior to final implementation on the Target System. This process was repeatable if EPROMS were used on the target card. This meant that all the software could be developed and tested before loading onto the target card (next section), thus cutting down on development time.

## The Target System Unit

This unit accommodates the target system which is used in the final implementation of the instrument. The SCIM 88 module (card) includes 256k battery backed static RAM, 512k Dynamic RAM, up to512k EPROM as well as the 80C188 processor (enhanced version of the 8088). The 80C188 processor is used as an embedded microcontroller (a small computer which supports all of the input and output functions as well as the software required for controlling the temperature). Communications is established between the host and the target system via the Sourceview software [1] and an SPCCOM STE Bus communications card which can be removed from the system on completion of development. This system also comprises its own power supply unit so that the system can be operated independently of the host and the optical unit.

Once satisfactory development has been established the target card can be removed and operated remotely e.g. on an independent backplane with any other necessary cards such as analogue or digital I/O and a small power supply unit. If desired the latter may be incorporated into the main Power supply unit of the final system.

## The Power Supply Unit

The rating of the power supply unit is largely dependent on the size of the Peltier Heat Pump of the optical unit. For the purpose of the current investigation a Heat Pump was used which could conduct a maximum of $\pm 3$ Amps from $\pm 6$ Volt supplies. Thus the power supplies required for this instrument were $\pm 5$ Volts, 12 Volts low current and $\pm 6$ Volts with a $\pm 3$ Amps capability. Clearly, the heat pumping capacity and hence the power supply requirements depends on the range of temperature differential between the copper block and the operating environment. It was necessary to be able to subject the control system to rather large changes in set point temperature for this investigation therefore, there was a need for a large pumping capacity.

## SYSTEM TESTING AND CALIBRATION

For the purposes of the present investigation the variables of interest were temperature of the copper block and the output intensity and wavelength of the Laser Diode. Therefore, it was necessary to be able to accurately monitor all of these variables during the course of the investigation.

The block temperature was measured using an LM135 temperature transducer which had an accuracy of better than 1% over its operating range after calibration.

The output intensity of the Laser Diode was monitored accurately using a Newport 815 Series Power Meter whose remote photodetector was mounted in close proximity to the Laser Diode Aperture in order to ensure good optical coupling. It was not possible to measure the absolute optical power output of the photodetector in this manner , but this was not required as a relative value of output was sufficient provided ht the optical system geometry was maintained constant throughout.

Changes in the output wavelength were measured as the spectral distribution of light over the entire wavelength range of the Laser Diode. Shifts in this distribution were evident when the block temperature was varied and these were measured on a Bentham (M1000) Monochromator which had the required resolving power (better than 0.1 nm). The system chosen for the measurement is shown in Figure 3.

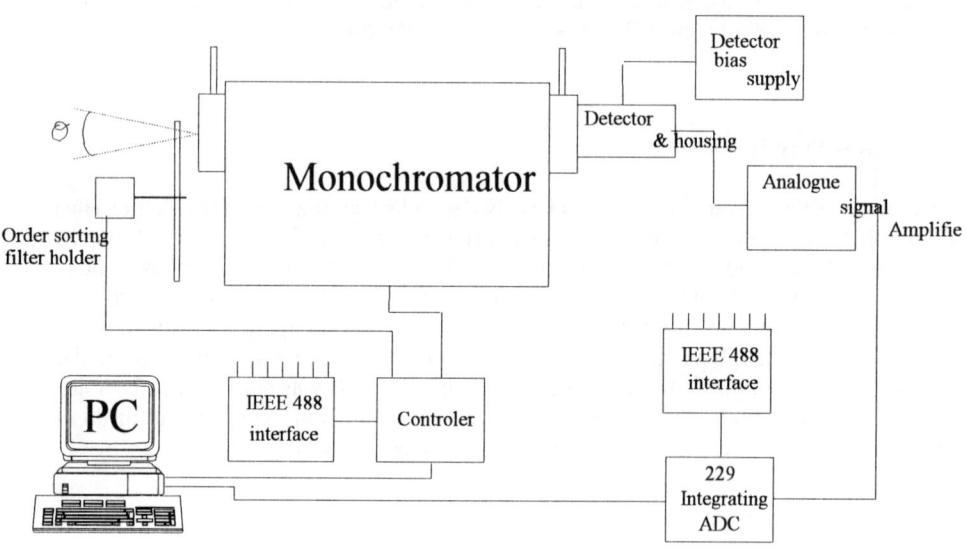

Figure 3. The System Configuration For Measuring Output Wavelength.

The monochromator used was capable of resolving wavelengths of 0.1 nm or better at the wavelength range of interest. The optical dispersion grating of the monochromator was driven via a stepper motor which was in turn controlled by a host PC. The host PC also performed the data acquisition in this case which comprised recording output intensity versus values of wavelength equivalent to just greater than the optical bandwidth of the Laser Diode (typically 5 nm) at a step interval of 0.02 nm.

Note that it is not possible to compare directly the values of output intensity in this case against those obtained previously with the Optical Power Meter. This is because the photodetector in the latter case has an integrating effect over the whole wavelength range of the output light whereas the monochromator makes discrete samples at 0.1 nm interval. However, provided the change in device temperature does not result in a serious change to the shape of the output spectrum it is possible to indirectly compare the change in the peak intensity value (at the centre wavelength) against that of the power meter case.

The monochromator was regularly calibrated using a stabilised white light source and a series of narrow band optical filters over a range of 200 nm centred about a wavelength of 780 nm.

## THE CONTROL SYSTEM

The temperature of the copper block was controlled by means of PID control. The schematic diagram of the system is shown in figure 4.

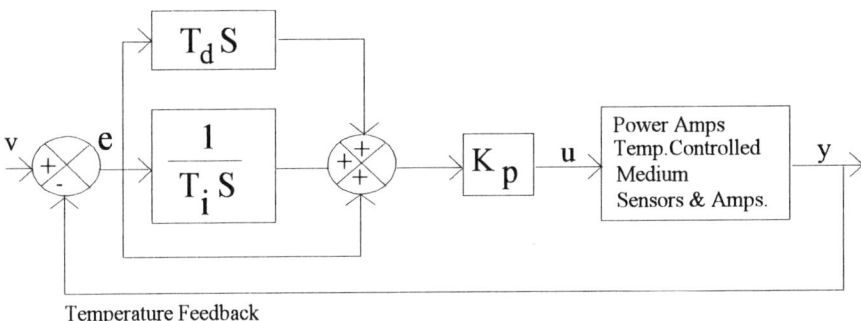

Figure 4. Schematic Representation of the Control System.

The time domain representation of the controller is described by the following equation

$$u(t) = Kp\left\{e(t) + \frac{1}{Ti}\int_{0}^{t}e(t)dt + Td\frac{de(t)}{dt}\right\}$$

The controller was tuned using the Ziegler Nichols approximation. Thus the values of the proportional, integral and derivative coefficients were based on the period of sustained oscillation in the critical gain regime. The results showing these responses are included in the results section.

Other combinations of controllers were used e.g. PI, PD, but the PID was found to be the optimum configuration for this controller. The transfer function of the PID controller is defined as follows :-

$$Gc(s) = Kp\left(1 + \frac{1}{Ti.s} + Td.s\right)$$

$K_p$ is the Proportional Gain
$T_i$ is the Integral Time
$T_d$ is the derivative Time

The time domain output of the PID is described as

$$U(t) = Kp\left\{e(t) + \frac{1}{Ti}\int e(t) + Td.\frac{de(t)}{dt}\right\}$$

e(t) is the input.

Two Ziegler Nichols methods are available for tuning and both were employed. They are summarised as follows:-

**Step Response Method**

From the step response of a first order system we can define the transfer function as

$$\frac{C(s)}{U(s)} = \frac{K.Exp(-L.s)}{T.s+1}$$

T is the time constant associated with the rise in the step response.
L is the delay time.

The following values of Gain and Time coefficients for a PID controller can be deduced using this method

| $K_p$ | $T_i$ | $T_d$ |
|---|---|---|
| $1.2\dfrac{T}{L}$ | 2L | 0.5L |

Table 1. Ziegler Nichols Step Response Tuning Values.

## Critical Gain Oscillation Method.

This is a more accurate method and makes use of the proportional term only. The time domain equation in this case is as follows :-

$$(t) = K_p + e(t)$$

A range of values of $K_p$ were investigated with the PID software running in proportional mode only. The critical gain value was established in this manner. The table of coefficients corresponding to this method is included below.

| $K_p$ | $T_i$ | $T_d$ |
|-------|-------|-------|
| $0.6K_{cr}$ | $0.5P_{cr}$ | $0.125P_{cr}$ |

Pcr is the period of oscillation with the critical value of gain, $K_{cr}$. In the case of the present system it was found that values for $K_{cr}$ and $P_{cr}$ of 10 and 40 seconds respectively were obtained.

It must be emphasised that the above two methods allow approximate values for the coefficients to be determined. It has also been assumed that the system is first order. In order to improve on this accuracy further work is required in identification, simulation and the application of more advanced control techniques which are beyond the scope of the work reported in this paper.

## EXPERIMENTAL RESULTS

The results of this investigation have been categorised as follows :-

- The Block Temperature versus time
- Output Intensity Versus Time
- Output Intensity Versus Wavelength

The results were recorded as outlined previously in the section describing the experimental arrangement. They are presented in Figures 5 to 8.

### Block Temperature Versus Time

Figure 5 Shows the response of the system to a step change in the set point temperature the set point temperature from, a value of 20.5 °C to 17.3 °C. The initial fall is followed by an overshoot to approximately 17 °C after which the temperature settles down to its set point value of 17.3 °C. The settling time in this instance is approximately 30 seconds. However the block temperature reached the overshoot value within a 20 second interval. The step response described in this case is characteristic of a second order system. For this reason it was necessary to carefully assess the values of the tuning coefficients derived in the previous section. However, the stability results indicate that the tuning coefficients were sufficiently accurate to provide good temperature control. There is however scope for further improvement of the control system by incorporating the techniques outlined in the previous section.

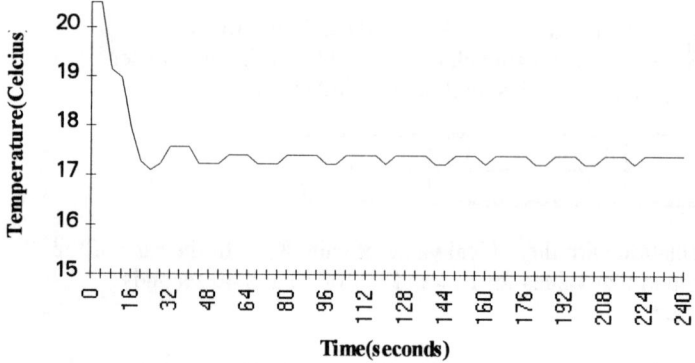

Figure 5. The variation of block temperature versus time.

## Output Intensity Versus Time

The temporal variation of output light intensity of the laser diode is shown in Figure 6. The values shown are in mW and correspond to the radiant power measured on a power meter as detailed in the section on the experimental arrangement. In this case the recording interval extends over 4 minutes and it is clear from this that the total intensity fluctuation is no greater than 0.2 mW. The average value of measured intensity over this period was 100.3 mW, thus the stability of the output intensity is 0.2 % or better.

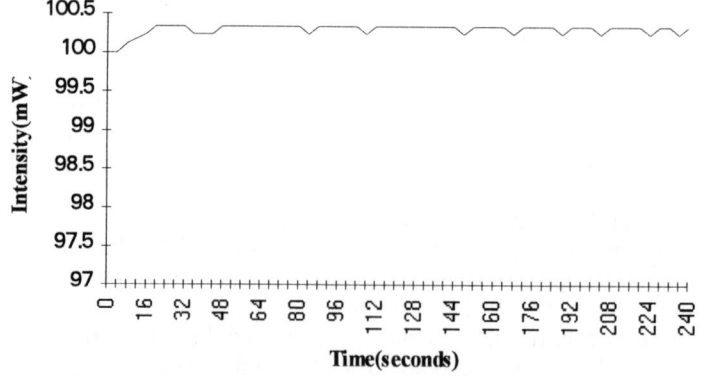

Figure 6. The variation of output intensity versus time.

The response of the system to a step change input is shown in Figure 7. This record includes the change in temperature as well as intensity when subject to an input step change in the set point value from 23.5 °C to 16.25 °C over an interval of 200 seconds. This result shows how the output intensity varies in sympathy with the block temperature.

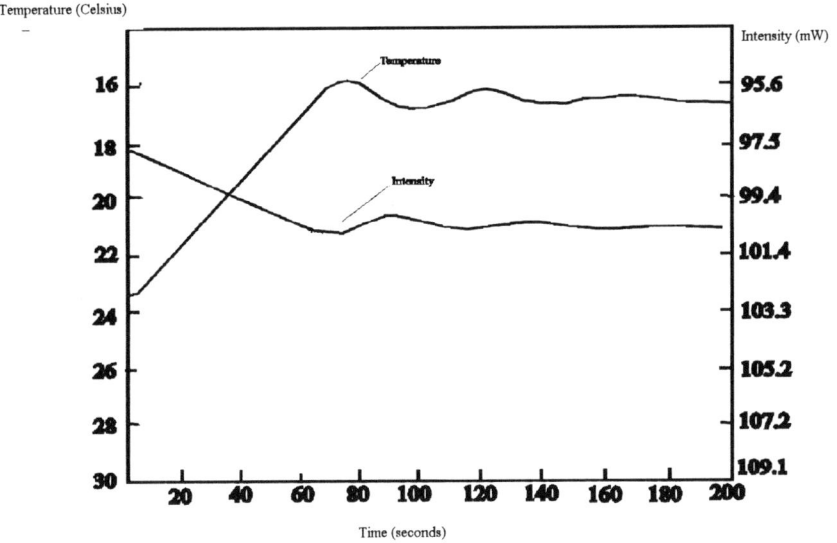

Figure 7. The recorded system step response

## Long Term Stability

The long term stability of the instrument is of paramount importance for applications such as down hole monitoring where periods of operation of hours to many days are often required. Therefore, it is essential that the source (or receiver) should be stable over such periods. Figure 8 represents a long term stability test where the temperature and intensity were monitored over a period of 500 minutes. During that monitoring period the maximum fluctuation in temperature was no greater than ±0.1 °C. Once again the total fluctuation in the measured output intensity was 0.2% of the average value (101 mW).

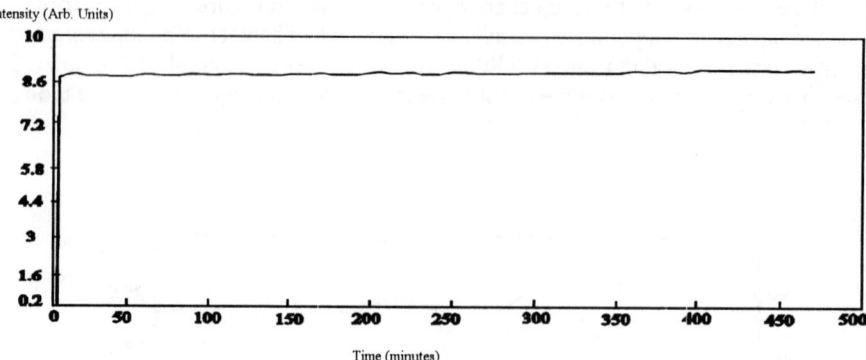

Figure 8(a)

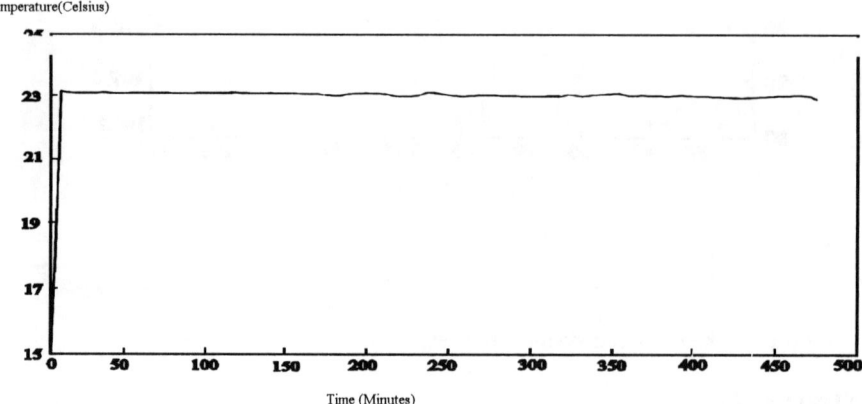

Figure 8(b)

Figure 8. Long term stability measurements for
      (a) Intensity
      (b) Temperature

Tests were also conducted in which the instrument was run continuously for several days.  The temperature and intensity were monitored at regular intervals and the fluctuations were not found  to be outside the limits already established in the results described above.

## Output Wavelength Versus Temperature

Output spectra from the laser diode for different block temperature values are shown in Figure 9.

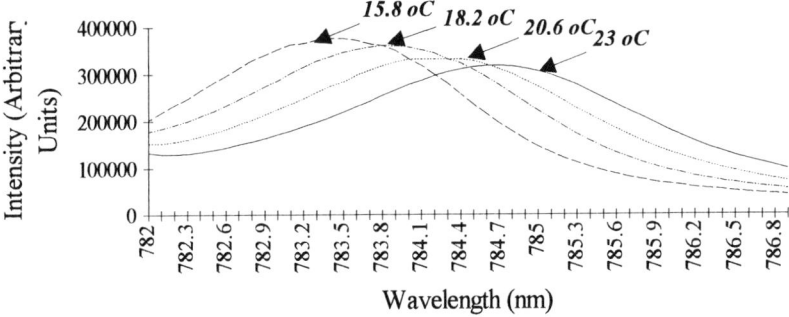

Figure 9. Measured Spectral Output of Laser Diode at Different
                Temperatures.

This clearly shows two major effects of altering the temperature.  They are as follows :-

•     A decrease in peak output intensity

•     An increase in the wavelength at which the peak intensity occurs.

The first of these effects has already been established earlier in the section on intensity measurements and the intensity changes are in line with those recorded previously(i.e. intensity decreases with increasing temperature).  However, it is dangerous to compare the peak intensity values directly to the values obtained with the photodetector due to different threshold levels of detection and the integrating effect (with wavelength) of the photodetector.

The second effect demonstrates that the output wavelength distribution of the laser diode is temperature dependent, a fact already well established for such devices.

However it is clear that a shift in the peak intensity wavelength value of 0.18 nm per $^{\circ}$C is in evidence. This change is reversible and reproducible. The total optical bandwidth of the laser diode is no greater than 3 nm, therefore the change in wavelength is significant if compared to this value. This change in intensity distribution has appeared as a smooth one and significantly the shape of the intensity-wavelength characteristic is not radically altered. It is known that this is true for all LEDs and low power laser diodes, but care must be taken when enforcing similar temperature changes on multimedia [11] higher power devices. In this case wavelength shifts result in mode hopping [12] which are totally discontinuous in their nature.

When the shift in wavelength is continuous and reversible it is possible to use this effect as a means of modulating the output wavelength of the device thus resulting in a wavelength tuneable source. Fluctuations in intensity when altering the temperature can be compensated by altering the drive current whilst still maintaining the same temperature. Although the range of wavelength tuning is somewhat limited in this case, it is useful in applications where such small changes can yield valuable information e.g. dual wavelength optical sensors.

## DISCUSSION AND CONCLUSIONS

The results described in the previous section have showed that a practical, rugged and versatile optical source suitable for illumination of undersea fibre optic systems is realisable. The instrument is particularly useful for fibre sensors owing to its high degree of stability (better than 0.2%) and flexibility in being able to house receivers as well as sources.

It is generally acknowledged that undersea sensor systems are required to survive extremely harsh environments and thus any source or detector located close to the point of measurement must be subject to the same conditions. The latter devices are normally shielded from their environment by extreme protective housings which often require comprehensive safety testing and approval etc. Although this is possible, it clearly adds to the manufacturing cost of the source as well as making routine maintenance difficult. In addition to this power must be supplied externally (via submarine cables) or internally by a battery which would require regular servicing.

The instrument developed during the present study is sufficiently versatile to accommodate a number of optical sources and detectors in any combination. Thus the source and detector can be physically located together or separately. Furthermore, the large choice of possible sources or detector allows the system designer to address a large number of sensing applications. For example, where large distances are involved between the point of measurement and the monitoring station it would be desirable to use a laser diode as a source, and an APD as a detector. Possibly the greatest advantage of using a fibre optic sensor in such situations is that the measurement is a passive one where only low power light levels (mW maximum) are present at the point of measurement.

The development of the instrument as an embedded microprocessor control system has resulted in a cost effective and versatile solution to an established problem associated with optical sensing i.e. long term stability. The fully developed version is a self contained unit which allows ease of transportation and satisfies the possible requirement of forming an integral part of a higher level monitoring system.

Results have shown that the utilisation of PID control for temperature stabilisation has resulted in a long term stability of the output light intensity of 0.2 % over a monitoring period of hundreds of minutes. It is possible to achieve further improvement in this figure using more sophisticated tuning methods and control techniques. This degree of stability is essential if accurate long term monitoring of undersea parameters is to be achieved by optical means. LEDs and Laser Diodes when operated in this manner can have greater lifetimes than standard white light sources. Values of up to twenty thousand hours are possible in some cases. Additionally, it is possible to operate the LEDs and Laser Diodes in chopped (switched) mode as well as Continuous Wave although this does result in shortening of the lifetime, particularly in the case of Laser Diodes.

It has been demonstrated that for the Laser Diode used the wavelength of peak intensity is dependent upon the device temperature. This is a well established fact and is also true for LEDs. The instrument developed for this study is capable of adjusting the device temperature in order to induce a wavelength change which is controllable and reversible. Thus the instrument it capable of wavelength modulation. The modulation range is dependent on the temperature range covered and the centre wavelength changes at a rate of 0.18% nm per $^{o}$C. This represents a significant shift when compared to the overall bandwidth of the laser diode ($\approx$3 nm). This ability of the instrument to vary the output wavelength is particularly useful for systems in which measurement is required at a multiplicity of wavelengths or a range of continuous ones. Although the instrument's settling time between wavelength changes is limited (of the order of seconds), this does not present a problem in applications where time constants are greater than this. The alternative to this would be modulation of the output wavelength by some external means such as acousto-optic modulation of an external diffraction grating [13]. Such devices tend to be prohibitively expensive (tens of thousands of pounds) and are more suitable for laboratory instrumentation.

Temperature controlled Laser Diodes have been used in conjunction with other optical sources such as Fibre Lasers. In this case fluorescence is induced in a specially doped fibre by 'pumping' the fibre with light from a Laser Diode source e.g. 830 nm for Neodynium doped fibre. Lasing action is achieved by enclosing both ends of the fibre with partially reflecting mirrors and laser light is emitted at 1080 nm. In such cases it is sometimes essential to maintain the peak wavelength of the Laser Diode constant since drift in this or the output intensity can result in reduced absorption of the pump light. The fibre laser source offers many attractive features for fibre optic sensor systems such as wide optical bandwidth and extremely good long term stability.

A robust and cost effective unit has been developed which is suitable for a wide range of optical based subsea and down-hole measurements. Many applications exist for the unit which include temperature, pressure, flow rate, displacement etc.. The hostility of the downhole environment requires the use of robust and intrinsically safe sensors which are readily fulfilled by certain optical fibre sensors. The unit in question is ideally suited for providing the illumination as well as light detection capability for such sensors and by use of optical fibres can be located remotely from the point of measurement e.g. at the well head. Furthermore, results reported in this research have shown that the unit is extremely stable against changes in its environment, thus providing the user with the capability of high accuracy measurement.

A high degree of versatility has been designed into the unit to allow interchangeability of the optical devices contained within it and further future improvements in controlling software to take place. This allows the user to tailor the system to their particular sensing requirement.

Future improvements to the developed instrument will include the application of more sophisticated techniques to the temperature control of the optical devices. Future control algorithms will include rule based logic i.e. Fuzzy Control. Also improvements will occur in the hardware such as greater resolution ADC/DAC (dependent on the signal to noise ratio of the measured variables) and the heat pump power handling (resulting in faster temperature changes) as well as the associated enhanced electronic drive capability.

A further application of this type of instrument would be to extend its use to Tungsten Halogen type 'White Light' sources. In this case it is only possible to modulate the driving voltage to compensate for changes in the output intensity. Therefore it would be necessary to include an optical feedback value as a control variable. In this case, a number of factors can be responsible for deterioration of light intensity but once these factors have been identified it will be possible to accurately control such sources in a cost effective manner.

## REFERENCES

[1]      Sourceview © Arcom Control Systems Ltd.

[2]      Dakin et al
         Novel Optical Fibre Hydrophone Using a Single Laser Source an Detector.
         Electronic Letters, Vol. 20, 1984, pp 51-53.

[3]      Rogers A.J.
         Distributed Optical Fibre Sensing.
         SPIE Vol. 1506. Micro-Optics II, 1991.

[4]      Black P.W.
         Undersea System Design Constraints.
         Proc. Soc. Photo Opt. and Instrum. Eng., SPIE 1314, pp112-115, 1990.

[5]     Dutta N.K. and Zipfel C.L.
        Reliability of Lasers and LEDs.
        in Miller S.E. and Kancinow (Eds.).
        Optical Fibre Telecommunications II, Academic Press, pp 671-677, 1988.

[6]     Zipfel C.Z., Chin A.K., Keramidas V.G. and Saul R.H.
        Reliability of DH $Ga_{1-x}Al_xA_S$ LEDs for Lightwave Communications.
        Proc. 19th Ann., IEEE Int. Reliab. Phys. Symp., pp 124-129, 1981.

[7]     Gwanmeh S.
        Implementation of a Fuzzy Logic Controller.
        MSc. Dissertation, Liverpool John Moores University, Sept. 1993

[8]     Mears R.J., Reekie L., Poole S.B., Payne D.N.
        Neodynium Doped Silica Single Mode Fibre Lasers.
        Electronics Letters, Volume 21, pp 738-740, 1985.

[9]     Jauncey I.M., Lin J.T., Reekie L., Mears R.J.
        An Efficient Diode Pumped CW and Q Switched Single Mode Fiber Laser.
        Electronics Letters, Volume 22, pp198-199, 1986.

[10]    Urquhart P.
        Review of Rare Earth Doped Fibre Lasers and Amplifiers.
        IEE Proc., Pt. J. Optoelectronics, 135 (6) pp, 385-407, 1988.

[11]    Senior J.M.
        Optical Fibre Communications : Principles and Practice.
        2nd. Ed., Prentice Hall International Ltd., pp 345-346, 1992

[12]    Yariv A.
        Introduction to Optical Elactronics
        Holt, Rinehart and Winston, 4th. Ed. 1991.

[13]    Coquin G, Cheung K.W. and Choy M.M.
        Single and Multiple Wavelength Operation of Acousto-Optically Tuned
        Semiconductor Lasers at 1.3 microns.
        Proc 11th IEEE Int. Semiconductor Laser Conf. (USA), pp130-131, 1988.

## GLOSSARY OF KEY TERMS USED

**LED**    Light Emitting Diode
**LD**     Laser Diode
**Multimode LD**.      Laser Diode which emits light in many launch modes.
**Photodector**.                    A device which delivers current when illuminated by light.
**PIN**    pin Photodiode. A low gain, high bandwidth photodetector
**APD**    Avalanche Photo Diode. A high gain photodetector.
**CW**     Continuous Wave. Mode of operation relating to optical sources where a continuous (DC) light output is obtained as oppose to pulsed output
**mW**     milliwatt - $10^{-3}$ W convenient unit of power for light tranmission /   detection.
**µW**     microwatt - $10^{-6}$ W convenient unit of power for light tranmission / detection.
**nm**     nanometre -$10^{-9}$ m convenient unit of distance for light wavelength measurement
**Monochromator**.      Instrument which accurately measures light intensity as a function of wavelength.
**Pigtailing**.      Technique in which a fibre is directly coupled to the end of a Laser Diode, LED or Photodetector.


**STE BUS**.      A standard cofiguration for the internal backplane connection of computers for industrial application.
**Backplane**.      A series of links or connections (usually on pcb) which interconnects internal computer cards or modules.  Includes data, address and control lines.
**ADC/DAC**.      Analogue digital converter / digital analogue converter module.
**RAM**      Random Access Memory
**EPROM**.      Erasable Read Only Memory.
**PSU**      Power Supply Unit
**80C188**.      Target processor used in this work
**SCIM88**.      Module card including 80C188
**SPCCOM**.      Communications module used in the instrumemt.
**SourceView**.      Software used for develpopment of embedded code.
**Borland C++**. High level development computer language.


**Peltier Heat Pump**.      Device which allows the transference of heat energy across it in either of two predetermined directions.
**PID**      A method of controlling a system which involves three terms i.e. Proportional, Integral and Derivative

# FIBRE OPTIC SYSTEM FOR DOWNHOLE MONITORING

B. Bjornstad
Alcatel Kabel Norge A/S
Ostre Aker vei 33
0509 Oslo
Norway

## ABSTRACT

The first fibre optic system for permanent monitoring of downhole pressure
and temperature has been successfully installed in an onshore gas well for
Nederlandse Aardolie Maatschappij B.V. (NAM) in the Sleen field in the
Netherlands. The system is design for 1000 bar pressure and 200°C
temperature, and has been fully qualified. A complete system has been
developed comprising passive gauge, downhole cable, penetrator unit for the
wellhead, wellhead and pod connectors (for subsea applications),
optoelectronics for sensor operation, and conversion and logging unit.

## INTRODUCTION

Early and reliable data of downhole pressure and temperature is required in
order to enable cost effective production, initially to refine and tune the
reservoir models and subsequently to optimise production and decide on
possible maintenance work.

Tools available to acquire downhole pressure and temperature information
can be divided in two main technology groups, i.e. wireline techniques and
permanent downhole installations, where each of the two approaches has its
specific characteristics [1]:

- Wireline techniques
  - Temporary installation, i.e. limited reliability
    requirement.
  - Special equipment and special service personal required for
    each survey.
  - Increased activity at the well imposes higher possibility of
    well damage during the survey.
  - High cost per survey.
- Permanent downhole installation
  - Long mission time, i.e. high reliability requirement.
  - Information available on request.
  - No involvement of special personnel at the well site, i.e.
    increase safety aspect.
  - No intervention cost for surveys.

Comparison of the listed characteristics shows the advantages of the
permanently installed systems. In order to make this option applicable, the
high reliability required for permanent installations should be realised in
a practical system.

*Volume 32: Subsea Control and Data Acquisition, 81–90.*
© 1994 *Society for Underwater Technology. Printed in the Netherlands.*

## FIELD EXPERIENCE WITH ELECTRICAL DHM SYSTEMS

Since 1987 specially developed electrical downhole monitoring (DHM) systems
became available and permanent installations were made in the North Sea.
Typically these systems include downhole installed sensors and electronics
(200 to 300 components), electric cabling, splices, wellhead penetration
and wellhead feed through, and remote electronics for powering and reading
the downhole gauge system.

Data has been accumulated for subsea installations in the North Sea, being
realised between 1987 and 1993, Fig. I. Some installations were made in
reservoirs with temperatures up to 120°C, but most were at temperatures
around 100°C. Results indicate that 20% to 25% of the systems failed
directly after the installation, whereas more than 50% failed within two
years; this failure probability is high in relation with the current
practices on planned intervention work, typically once every 5 years.

From those installations were failure modes could be determined, it
appeared that roughly 1/3 was related to the downhole sensors and
electronics, 1/3 was related to the cable and 1/3 to connections.

For future DHM applications two additional aspects will affect the
reliability requirement for permanently installed systems and reduce the
applicability of electrical DHM systems, i.e.

-        Extension of the period between interventions to up to 10 years,
         enlarging the discrepancy between reliability target and the
         current reliability performance of electrical systems.

-        New applications in deeper reservoirs will have higher pressure
         and higher temperature characteristics.

The high temperatures in future applications impose a severe problem on the
use of electrical DHM systems. Standard electronic components used in
today's downhole gauges can operate normally at temperatures up to 90°C.

Only special 'MIL-spec' components can be used for applications up to 120°C
continuously and up to 150°C on short term bases. A very limited range of
components is available as high temperature electronics, however,
insufficient in range to realise the complexity of the downhole circuitry.

The reliability of permanently installed electrical DHM systems is
insufficient for current applications and will be grossly insufficient for
future applications with high pressure and temperature characteristics and
longer periods between interventions. In that light, a new and diverse
technology approach to the problem is made to realise the high reliability
targets.

Statoil have made a strategic investment in 40 permanently installed
electrical systems in the Gullfaks and Veslefrikk fields in the Norwegian
sector [2]. This investment was based on a threefold incentive:

-        The need for enhanced reservoir description, especially important
         during the initial production phase.

-        Increased production resulting from a combination of less
         downtime for data acquisition and optimisation of reservoir and
         well management.

-        Safety and operational considerations.

An economic analysis has been performed of this investment showing a
payback period of less than one year.

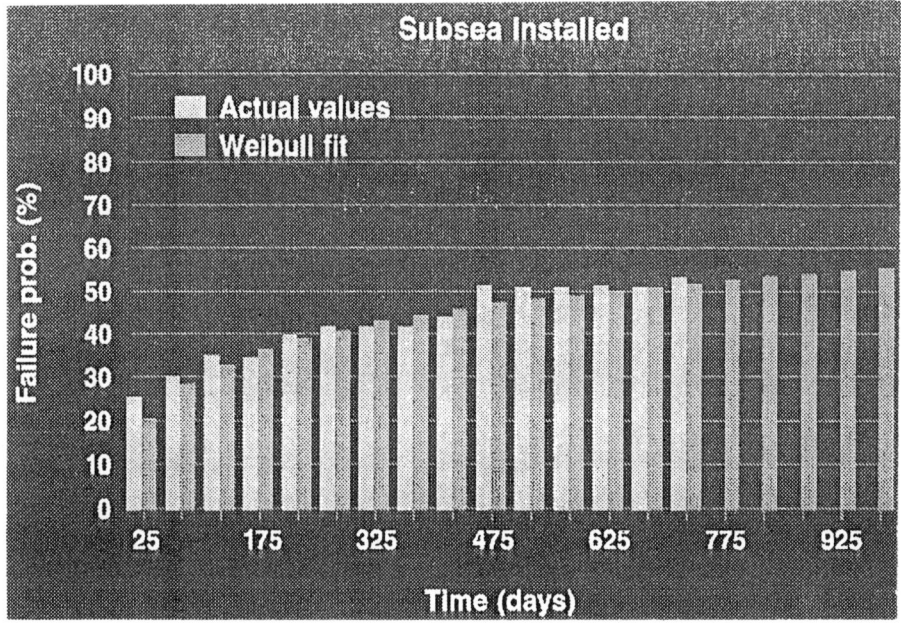

Fig. 1.  Reliability Electric DHM Systems

Petrobras plan to invest in a $19 million programme to install 65 permanent
gauges in subsea wells by 1999, and expect to save at least $79 million on
this investment by eliminating 260 pressure surveys that would have been
performed with drilling rigs [3].

Both the examples above show the economic advantages of investing in
permanently installed DHM systems. This advantage will increase
dramatically when a new generation of more reliable DHM system is made
available.

## FOWM

The Fibre Optic Well Monitoring (FOWM) system is a new type of permanently
installed DHM system developed by Alcatel Kabel Norge under contract by
Norske Shell, BP Norway, Norsk Hydro, and Norwegian Government, with Norske
Shell as the main sponsor [4]. In contrast to the currently available
electrical systems, the FOWM system is based on an optical sensor system
concept, integrating simple passive silicon resonators with optical
communication [5].

| Pressure range | 1 - 1000 bar |
|---|---|
| Temperature range | 20 - 200°C |
| Accuracy | < 0.4 bar |
| Resolution | < 0.1 bar |
| Time response | 1 bar/second |
| Vibration | 20 g, 30 - 100 Hz |
| Shock | 190 g, 9 ms |
| Useful life | ≥ 5 years, 90% probability |

Fig. 2.   System Specifications

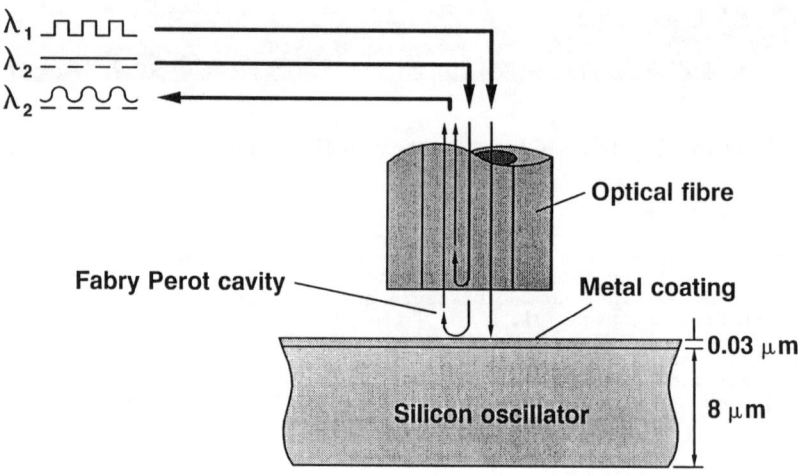

Fig. 3.   Principle of Operation of the Silicon Renonator

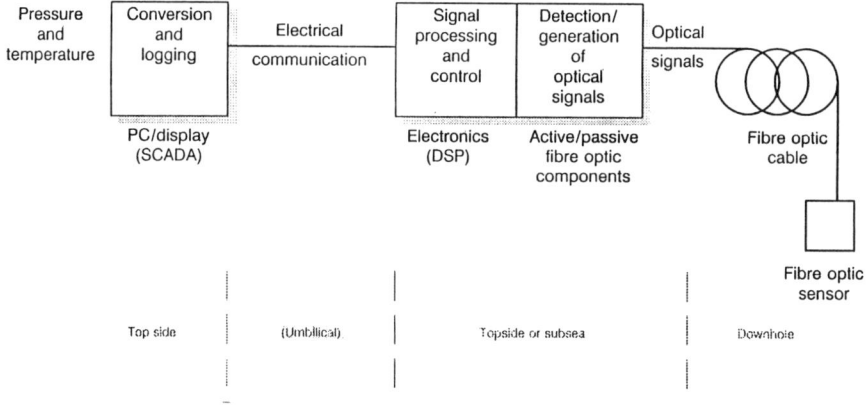

Fig. 4.   FOWM System

Not only the high reliability requirement for permanent installations, but also the anticipated high pressure, high temperature applications are covered in this concept, what is reflected in the main design parameters, Fig. II:

- Maximum reservoir pressure:              1000 bar
- Maximum reservoir temperature:           200°C
- Reliability:                             90% probability to remain functional for 5 years.

Main elements that contribute to the high reliability potential of the system include:

- Reduction of the number of downhole components; only passive pressure and temperature sensors are situated downhole.

- Complete elimination of downhole electronics; all required optoelectronics are platform based, simplifying access for possible maintenance.

- Use of a sensor concept which by design does not need additional hardware downhole for temperature compensation.

- Exploitation of a fault tolerant fibre optic connector concept for application in wet environments.

- Exploitation of a fibre optic communication channel, insensitive to the equivalent of the electrical short circuit failure mode.

- Slim cable design with laser welded steel tubes of "unlimited length", minimising the risk of cable damage during installation and operation.

The sensor element is based on optical excitation and interrogation of a micromachined silicon oscillator, Fig. III. The oscillator is excited into resonance by modulating laser light at one wavelength transmitted through

an optical fibre to the sensor. The light is absorbed by the oscillator
which induces thermal gradients and thereby vibration. Continuous light at
an other wavelength for interrogation is transmitted through the same
fibre. The detection principle is based on the interference between light
reflected from the fibre end and light reflected from the oscillator
surface. The resonance of the oscillator is determined by the pressure and
temperature downhole. The principle of phase modulation makes the system
practically insensitive to fluctuations in signal attenuation, and allows
for a large separation in distance, more than 25 km, between the gauge and
the optoelectronics.

The main building blocks of the FOWM system are, Fig. IV:

-   Downhole gauge; contains both the optical pressure and
    temperature sensor elements.

-   Fibre optic cable; transmits optical signals downhole.

-   Optoelectronics with lasers and detectors; runs the sensors, as
    well as performs optical-electrical signal conversion.

-   Digital signal processing unit; controls the lasers and
    communicates with the conversion and logging unit (CLU).

-   CLU; converts from frequency information to pressure and
    temperature data based on sensor characterisation curves loaded
    in software. The CLU also presents it on the screen and logs the
    data to file.


## FOWM Components

From the early start of the project, the reliability requirement was a
major design criterion, i.e. system components have been designed for
application in a permanently installed system with the related high
reliability requirement.

The FOWM gauge contains one pressure sensor and one temperature sensor, and
has a slim design measuring only 1 inch in diameter, which is an advantage
especially in slim bore wells.

The laser welding technique utilized in steel tubes containing the fibres
in the cable enable manufacture of "unlimited" lengths of cable without
factory splices. The current standard cable length is 10 km. This is a
great advantage considering that electrical cable typically fail in the
factory splices.

The optical fibre penetrators use slim 1/8 inch tubes, and serve as
pressure barriers down to the fibre level in the wellhead. Units have been
long term tested for 6 months at 175°C and 800 bar without leakage.

The CLU controls the system and presents and logs pressure and temperature
data. The core unit, optoelectronics, has been realized on only 3 standard
Europe-cards, and can be mounted in control pod or rack. The system has
standard RS232 interface and can be integrated with control systems, or
come as a stand-alone unit with data displayed on a rack mountable PC.

Subsea completions require wet mateable connectors, and units for both
wellhead and pod applications have been developed and tested [6]. The units

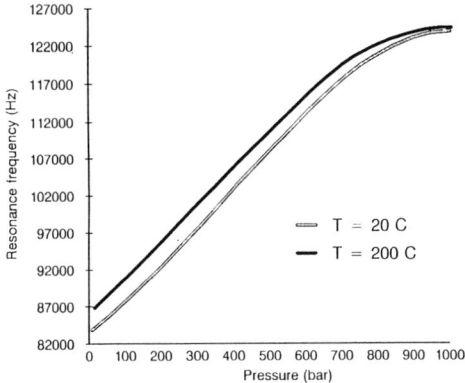

Fig. 5. Frequency Response of the FOWM Gauge

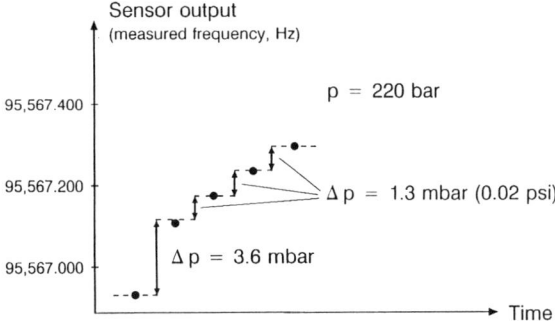

Fig. 6. Resolution of the FOWM Gauge

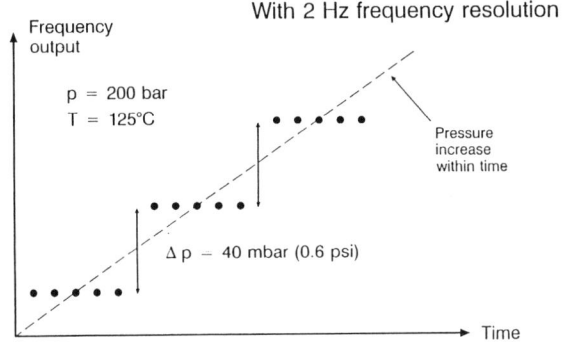

Fig. 7.  Resolution of the FOWM System

Trade off between:
▼ Resolution
▼ Time Response
▼ Signal Amplitude

|                    | Time response          | Resolution              |
|--------------------|------------------------|-------------------------|
| "Practical" limits | 30 bar/s               | 0.08 bar   (1.2 psi)    |
|                    | 0.3 bar/s              | 0.004 bar  (0.06 psi)   |
| NAM installation   | 1.5 bar/s              | 0.04 bar   (0.6 psi)    |

Fig. 8.

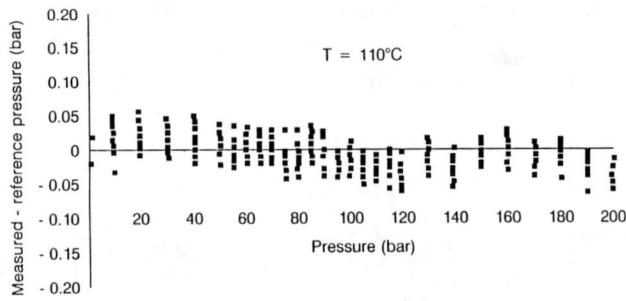

Fig. 9.   Pressure Readout of the FOWM System Compared
          to a Reference Pressure

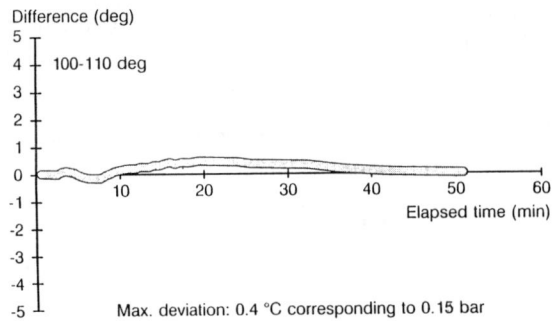

Fig. 10.   Temperature Measured with Pressure and Temperature
           Sensor (THB2)

have less than 0.5 dB attenuation after series of 10 matings/dematings in contaminated water (1% barytt), under pressure (200 bar), and after being covered in sand (10 cm of sand flushed away before mating).

## System Performance

Since the project start in 1986, various development and test phases have been conducted, including full qualification testing with results according to specifications.

The gauge has a linear frequency response both with respect to pressure and temperature loads, which makes it easy to implement the conversion algorithms in software, Fig. V. The pressure response is 50 Hz/bar and the temperature response is 15 Hz/°C.

The resolution of the gauge itself has been measured to be 1.3 mbar, Fig. VI, while the resolution of the complete system is determined by the frequency resolution of the electronics of 2 Hz, or the equivalent of 40 mbar, Fig. VII.

The system has a large degree of flexibility, and there is a trade off between resolution and time response (changes the system can follow and stay within specifications), Fig. VIII. The user's needs will vary during the life of a well, and the user can in a certain time want to follow large pressure transients. FOWM can follow as much as 30 bar/s, which results in a resolution of 0.08 bar. The user may in an other phase of the well's life cycle want emphasis on resolution, and 0.004 bar is possible, corresponding to a time response of 0.3 bar/s.

The accuracy of the system has been studied by running FOWM in parallel to an accurate reference sensor, Fig. IX. Deviations of less than 0.05 bar have been observed.

The FOWM system use the temperature sensor for compensation of the pressure measurement. The sensor elements are positioned approximately 5 cm apart resulting in a small error in the pressure reading during temperature transients. The maximum error reading is only 0.15 bar during the 15 minutes interval it takes for the two sensor elements to read the same temperature after exposure to a 10°C increase in temperature load, Fig. X.

The first system was successfully installed in an onshore gas well in the Sleen field in the Netherlands for NAM in October 1993. The gauge is set at approximately 1800 m, and the downhole conditions are 122 bar and 82°C. The system is performing within specifications.

The offshore debut will be made in the Gyda field, operated by BP, on the Norwegian continental shelf in April 1994. This installation is of particular interest having downhole pressure and temperature of respectively 550 bar and temperature above 160°C.

## CONCLUSION

–    The first fiber optic well monitoring system has been installed
     successfully.

–    The FOWM system has been qualified for applications upto 200°C and
     1000 bar.

–    Test results verify that the FOWM system meets the target
     specification in terms of resolution, accuracy, and time response.

–    The availability of the optical connector makes subsea completions
     possible.

–    The FOWM system has been commercially available since November 1993.

[1]    P. Eigenraam (1992), "Development of a reliable, permanently
       installed downhole monitoring system", Production and Process
       Technology Conference, The Hague

[2]    T. Unneland, "Permanent Downhole Gauges Used in Reservoir Management
       of Complex North Sea Oil Fields", SPE 26781, Offshore Europe,
       Aberdeen, 7-10 September 1993

[3]    C. Fox, "Permanent gauges go in at the deep end", Offshore Engineer,
       February 1993, pp 21-23

[4]    B. Bjørnstad et al, Fibre Optic Well Monitoring System", SPE 23147,
       Offshore Europe, Aberdeen, 3-6 September 1991, pp 425-432

[5]    T. Kvisterøy et al, "Optically excited silicon sensor for permanently
       installed downhole pressure monitoring application", Sensors and
       Actuators A, 31 (1992), pp 164-167

[6]    A. Berg, "Underwater mateable connector with single-mode fibers",
       Fiber Optic Components and Reliability, SPIE, Boston, 3-6 September
       1991

# THE DEVELOPMENT AND UTILISATION OF THE **FLUID-FILLED ELECTRICAL CONNECTOR**

D.B. PYE
Technical Sales Manager
Tronic Electronic Services Limited
Sandside Road,
ULVERSTON.
Cumbria.  LA12 9EF. England.

ABSTRACT

The electrical connector is rapidly becoming a critical component in Subsea Control Systems.  In recent years the advent of the Fluid Filled connector, in all its various guises, has greatly improved the reliability of the device and allowed its use in more and more critical applications.

This paper seeks to provide an insight into the connector, its design, development, testing and implementation in a number of applications and also to explain the functioning of one particular type.

A brief history of the evolution of the subsea electrical connector in general is also provided along with descriptions of three variations of the basic fluid-filled principle and termination methods.

The paper also calls for continued industry involvement in the further progression of the product.

HISTORY

The use of electrical equipment subsea is now an essential factor in both Commercial and Military activities.  The earliest "diving boats", dating back to Robert Fulton's NAUTILUS of 1798, were obviously purely mechanical devices having only two hull penetrations, namely those for propulsion and steering.

The earliest use of electrics subsea appears to have been in 1881 when the first practical electrically driven submarine was developed by the French engineer GOUBET.

*Volume 32: Subsea Control and Data Acquisition*, 91–111.
© 1994 *Society for Underwater Technology. Printed in the Netherlands.*

In these early vehicles however, sea water and electricity were
kept strictly apart and no hull penetrations for cables were
required.   This changed in 1908 when a submarine equipped with
telephone and lights was produced, again by a Frenchman, Abbe
Raoul.   The penetrations here were most likely simple stuffing
glands which could give no protection in the event of cable
damage.

By far the greatest impetus to any technological development is
given by the onset of war, but, whilst the submarine itself was
extensively developed during the First World War, the greatest
leap forward in underwater instrumentation came during the Second
with the introduction of Sonar ("ASDIC") systems and
sophisticated listening equipment such as Hydrophones.

These detection systems could obviously be used in static
applications as well as on submarines and their deployment in
defensive roles necessitated the development of cable and
penetration systems suitable for long term immersion in very
hostile conditions.

The need for reliable underwater electrical systems was obviously
now a high priority.


**EARLY SOLUTIONS**

The simplest method of penetrating the wall of a pressure vessel
is by stuffing gland.   This is suitable only for modest pressures
and, because of extrusion, not for extended periods.   Most
significantly however, particularly for manned vessels, no
protection against water ingress through a damaged cable is
provided.   Thus the severing of a cable, a not infrequent event
in the early days, would almost certainly result in a flooded
vessel.

The initial solution to this problem came with the development of
the Water Blocked Penetrator.   This device very effectively
prevented water ingress through damaged cables but was unable to
prevent the loss of electrical integrity of all circuits even in
the event of only one being damaged.

The earliest connectors also prevented water ingress but were of
questionable reliability, almost invariably being seen as very
much the weak link in any electrical system.   Penetrators were
used wherever possible or large stocks of self-amalgamating tape
were kept to hand !

**DEVELOPMENT STAGES**

Three distinct phases of development are in evidence in Subsea
Electrical Connectors :

1.   Rubber Moulded
2.   Metal Shell
3.   Fluid Filled

Rubber Moulded connectors first appeared in the 1940's and some
of the basic designs are still widely used today.  These are
simple and effective and provide very cheap solutions.  Most
however are not very robust, cannot be assembled subsea and
suffer from deterioration of electrical properties over quite
modest periods of immersion.

Metal Shell connectors address the problem of robustness very
effectively and some designs are mateable subsea.  They still
suffer however from long-term deterioration both of contacts
through corrosion and of electrical properties through water
penetration.

The present Fluid-filled designs have gone a long way towards the
ideal solution of ease of handling and long-term integrity (both
of structure and electrical properties) by once again separating
the critical electrical components from contact with sea water.
Additionally, the use of pressure-compensation has increased the
design life of the equipment by removing stresses from critical
components such as seals.

The exact method of containing the contacts in an insulating
medium varies from design to design but three main solutions have
arisen, notably:

a.   **Oil Flush:** This relies on the replacement of sea water with
     oil by flushing after engagement.

b.   **Sphincter gland.** Receptacles are contained in a grease
     filled chamber and the pin enters through a hole which is
     enlarged by the pin itself.

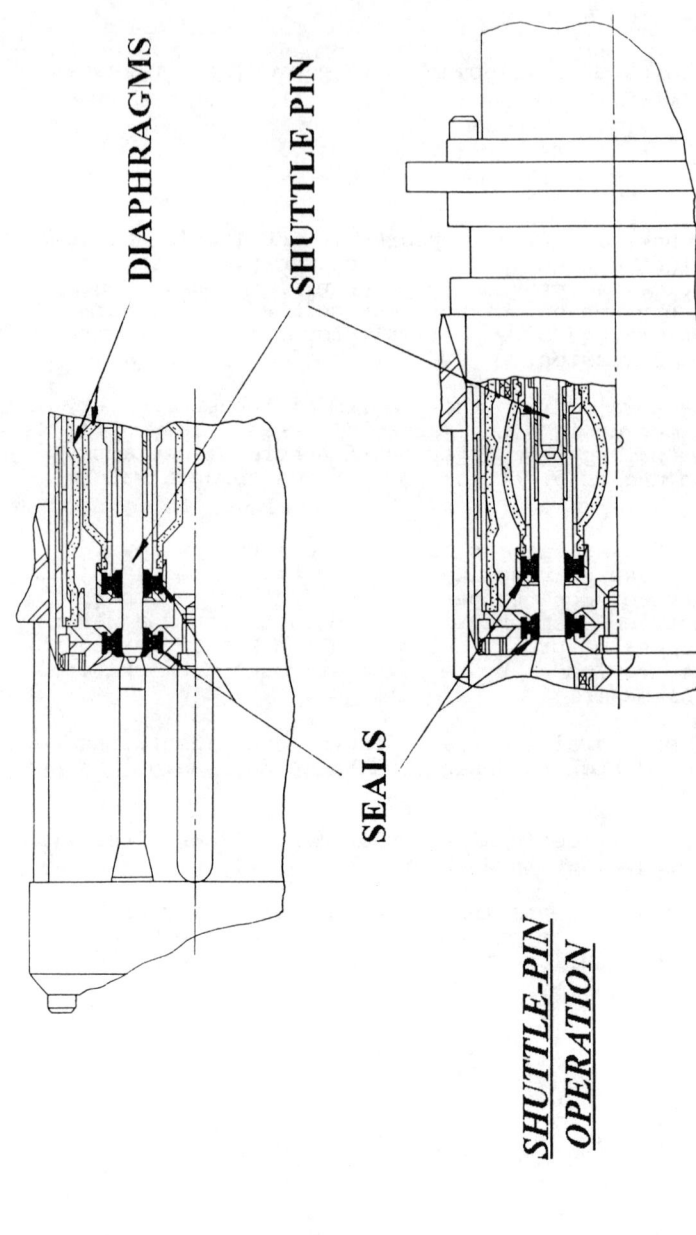

DIAPHRAGMS

SHUTTLE PIN

SEALS

*SHUTTLE-PIN OPERATION*

*FIGURE 1*

c.    **Shuttle-pin.** Receptacles are contained within an oil-filled chamber which is sealed when disconnected, by a non - conducting pin.  On assembly this shuttle is displaced by the pin contact which enters the oil-filled chamber and makes contact with the receptacle therein.

All types have advantages and disadvantages, the most significant being:

**Oil flush:**  This is the simplest of all the solutions but does require additional tooling to provide the flushing action. Electrical insulation properties prior to flushing are virtually non-existent.

**Sphincter gland.** Allows all contacts to be mounted on one pin, thus eliminating the need for keyway - a very important feature in remote connection.  Its integrity however is dependant on the action of the gland which has to open and close reliably.  Water ingress also is likely to affect all contacts of the connector.

**Shuttle pin.** Provides easily for multiple levels of integrity and prevents ingress affecting all contacts.  To its detriment, this design is the most complicated in terms of numbers of components. This problem can however be alleviated by having the most complex half retrievable. (See Fig. 1)

## CHANGING REQUIREMENTS

The most significant change in the design criteria of Subsea Electrical Connectors has been the requirement to maintain electrical properties (and obviously structural integrity) over periods of 20 years.  This period has further increased to 50 years in some recent applications.

To obtain this performance from a device that has to be assembled in a very hostile environment, either by a heavily-suited Diver or by a quite non-dextrous machine, was a significant challenge to designers.

Prior to the use of Subsea Production facilities, connections could be made and tested at the surface before deployment.  The increased exploitation of marginal resources has inevitably resulted in the requirement for these tasks to be carried out at great depths, initially by diver but lately by Remotely Operated Vehicles (ROV's) or Atmospheric Diving Suits (ADS).  This again changed the connector design criteria by imposing new handling problems.  A diver has the advantage of both "feel" and local 3D vision when assembling precise items such as connectors, whereas an ROV does not.  The use of ADS goes part way to solving the problem of 3D vision but retains some of the divers limitations as to endurance.

Another significant requirement which, until the development of
the Fluid-filled connector, could only be met by the use of
Inductive Couplers, is that of live mate/de-mate. This
capability is seen as advantageous in that equipment can be
removed for repair or replacement without the need to shut down a
facility completely.

Whilst the connectors ability to meet this requirement has been
proven under test conditions, it remains debatable whether a
Diver would be confident enough to perform the task despite these
results !

The new era of Marginal Field Development thus necessitated a
total rethink on the part of the connector designers in order to
meet these new requirements.

## DESIGN CONSIDERATIONS

As mentioned above, a number of new criteria have been imposed on
connector designers in the past few years and, as would be
expected, a number of differing solutions have arisen. One
characteristic has however been common to all - that of
containing the contacts within a controlled environment.

The main design considerations are discussed here in more detail:

### Long-term maintenance of electrical properties:

As noted previously, this has been achieved mainly be ensuring
that the contacts are maintained in a "controlled environment",
i.e. Oil or Grease, throughout their lifetime. In addition to
this, particular attention has been paid to the results of water
ingress through the connector to cable interface. This area is,
in my opinion, of equal if not greater importance to the
integrity of the connector design as a whole. No matter how
advanced the interconnection system may be, it can still be
rendered ineffective by the method of connection to the cable.

Pressure-balanced termination techniques are used here as well to
reduce the effects of pressure over long periods and these can be
backed-up by internal boot arrangements to minimise the effects
of water ingress through one core alone.

Termination is carried out by three basic methods:

### a.    Moulding

The simplest, but not now the cheapest method of termination,
moulding has advantages for easily and regularly retrieved
equipment. Developments are however taking place that may make
the method suitable for longer periods of immersion.

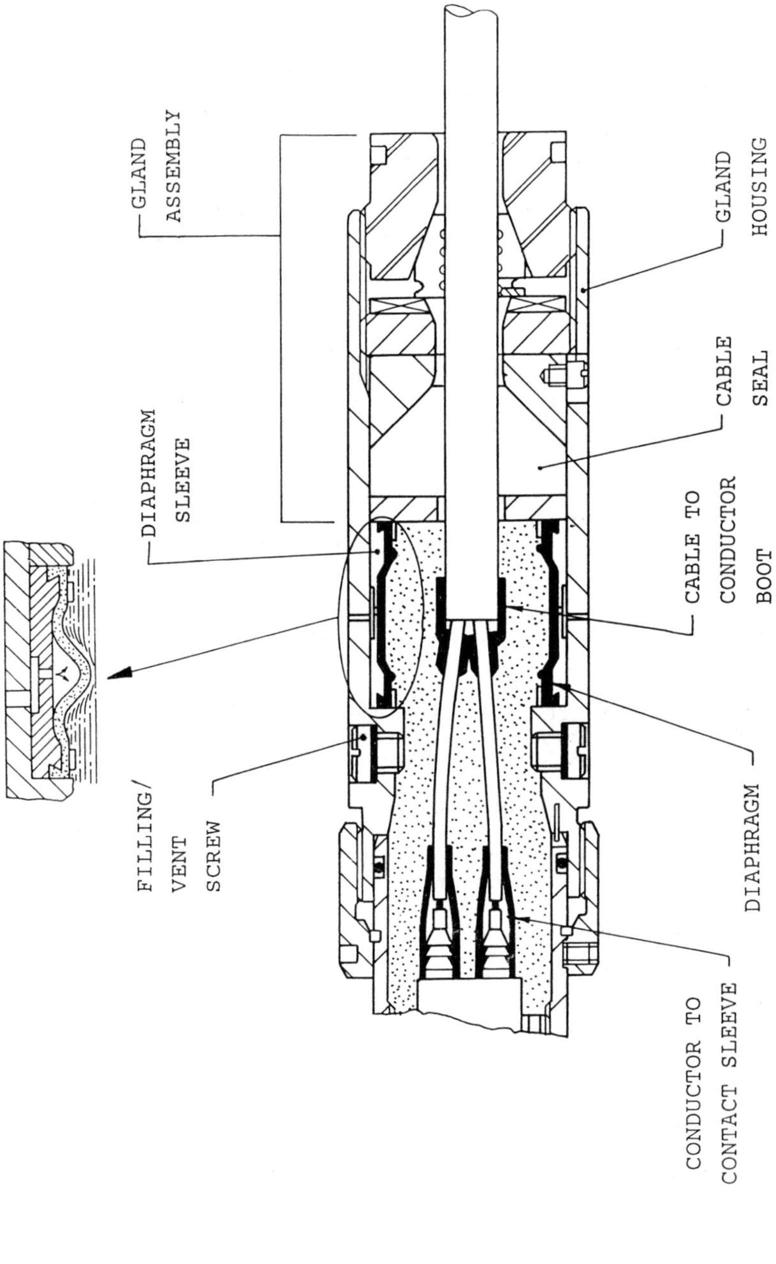

GLAND ASSEMBLY

GLAND HOUSING

DIAPHRAGM SLEEVE

CABLE SEAL

CABLE TO CONDUCTOR BOOT

FILLING/VENT SCREW

DIAPHRAGM

CONDUCTOR TO CONTACT SLEEVE

TYPICAL FIELD INSTALLABLE GLAND ASSEMBLY

## FIGURE 2

b.    **Pressure balance gland**

Based on the well-proven "stuffing-gland design" but with many
detail refinements such as the introduction of flexible
diaphragms and internal boot seals.  The use of pressure
balancing removes the stresses from seals thus making the
assembly suited to long-term departments.

Gel - or oil-filling gives and additional level of protection
against water ingress and allows the termination to be remade in
the event of cable damage. (See Fig. 2)

c.    **Oil-tube**

Records show the earliest use of this method dating back to 1930
when Dr. William Beebe used a flexible conduit to contain
Telephone and Electrical cables on the world's first practical
Bathysphere.  Whether the conduit was oil-filled or not is
unclear - most likely not.

Things have obviously progressed from that time both with the
construction of the tubing and its termination to the connector
or penetrator which may be in the form of a standard swaged
hydraulic fitting.

The use of this type of termination allows easy assembly by
personnel with only minimal training and facilitates fast repair
in the field.

It is anticipated that increased use of the oil-tube terminations
will occur and that developments in the tubing design will still
further enhance its reliability and flexibility.

Both the area around the contacts AND the cable termination
benefit from pressure balancing and these secondary and tertiary
levels of protection are seen as providing the solution to long-
term integrity of the termination in general.

**Maintenance of structural integrity**

Good electrical properties are irrelevant if the connector does
not prevent water ingress to the vessel on which it is mounted
for its design life.  Once again, the use of multiple barriers
and pressure compensation are much in evidence in the design.

One of the most difficult operating criteria for designers to
meet is that, unlike most electrical connectors, these have to be
capable of being easily disconnected after years of exposure to
corrosive elements, sand, silt etc.  This is far more difficult
than designing for frequent disconnection.

Where disconnection is impractical it is therefore still better
to design in a Penetrator which, in spite of the greatly improved
reliability of the latest generation of connectors, removes one
mode of failure from the equation.

Material selection and careful design of latching mechanisms have
thus become of primary importance.

### Handling

Unlike topside-mateable connectors, which can be assembled in
reasonable environmental conditions with perfect visibility,
Subsea Mateables have to be extremely tolerant to rough handling.
This has caused the introduction of many new features to the
designs ;

a.   **Keyways**

These have become essential to the ease of assembly of the mating
halves and have to be both robust and to provide a tapered entry
giving progressively more accurate alignment as the halves
engage.  They must totally prevent contact between pins and
sockets prior to full keyway engagement ("scoop proof") and be
resistant to contamination by silt or similar deposits.

b.   **Latching**

The method of holding the two connector halves together varies
significantly between Diver and ROV mateable applications.  In
both cases however the design philosophy is identical - "keep it
simple".
This is of course the key to all successful designs no matter
what the field of expertise.  It is important also to seriously
consider the long-term effects of corrosion, marine growth and
contamination on the mechanism.

c.   **Compliance**

Particularly in the case of ROV/ADS intervention, it is important
to minimise the effect of both the lack of feel, reduced
visibility and reduced dexterity of the handling devices.  Spring
mountings are provided to give lateral, longitudinal and a degree
of rotational movement to the receptacle.

## d.    Consistent Insulation and Contact Resistances

This is where the Fluid-filled connector exhibits its greatest
advantages in the remote assembly scenario.  Earlier designs of
connector required connection to pre-determined minimum torque
levels to achieve satisfactory Insulation Resistance levels and
Contact Resistance could be varied by contaminants.  This is
totally eliminated by the Fluid-filled connector which provides
multiple seals and chambers for high IR and wiping seals to
ensure clean contacts.  Additionally, a degree of longitudinal
tolerance is also incorporated to allow a small amount of
movement between connector halves without compromising these
essential properties.

## MATERIAL SELECTION

The development over the last 10 years or so of new, corrosion
resistant alloys, has made the connector designers job both
easier and harder !  Materials are now available that promise
improved performance over the immersion periods specified but
their actual performance in service is still a matter for debate.
Two approaches can be taken, either to use relatively
straightforward materials such as 316L  with reliance on cathodic
protection, or to go for more exotic materials such as Inconel
and Titanium and isolate these from protection systems.
The arguments are involved and heated !

However, it can be said with confidence that materials with much
improved resistance to corrosion and marine growth are readily
available to designers.  Not only the long term performance of
metallic components but also that of the Plastics and Elastomers
used have to be carefully evaluated.  Again, significant
improvements have been made in recent years and the choice has
been increased enormously.  It will, in my opinion, still be a
number of years before extensive use of plastics to replace
metallic components is achieved.

## TESTING AND DEVELOPMENT STAGES

The testing of the finished design is obviously of prime
importance in building confidence in prospective users.  These
connectors have become more and more essential components in the
function of subsea systems.  Their use is being extended from
simple instrumentation applications to the main power
terminations on the umbilical cable itself, hence the necessity
for complete reliability.

Again we here encounter the problem of the virtual impossibility of simulating, with any degree of accuracy, the effects of 20 years plus of continuous immersion in a most hostile environment. Whilst Accelerated Lifetime Testing can go some way towards this, there is absolutely no substitute for in-service records.   To best simulate the conditions in which the connector will have to operate, extensive test programmes have been laid down in consultation with the potential customer.  Some details of these follows:

a.   **BP Sunbury**

100 manual mate/de-mate cycles carried out in a Turbid Tank at atmospheric pressure.  Insulation resistance figures checked after every 10 cycles, exceeding 100 M Ohms on completion.  This test was carried out on a number of different connector designs.

b.   **National Engineering Laboratory**

This was a theoretical Failure Mode and Effect Analysis commissioned by OCCIDENTAL. Dimensional and material changes were recommended and incorporated, and further tests were suggested.

c.   **Norsk Hydro**

A number of connectors were supplied to the Norsk Hydro Research centre at Porsgrunn for long term investigation under, as near as possible, operational conditions.

These were all 4 way connectors and 3 pairs were retained at Porsgrunn whilst a fourth was sent to NUTEC at Bergen for shallow water deployment.

Initial pressure tests at Porsgrunn revealed some design deficiencies related to non-metallic components which were resolved prior to commencement of the long-term testing.  Tests at Porsgrunn were terminated in March 1993 with the longest period of test being one of 5 years 3 months.  Insulation resistance on completion was of the order 500 M Ohms and 50 mating cycles had been carried out.  In November, 1992, one of the original connectors was replaced with an example of the  MINI CE which is still under test.

d.   **Tests on Snamprogetti and S.I.S.L. variants**

High-power connectors are finding increasing interest in their deployment subsea.  Connectors rated at 400 Amp/11000 v and 125 Amp/3300 V have undergone extensive test programmes during the past few years.  These tests pointed out various necessary refinements and have resulted in products that will find many applications in the near future.

3.    **In-house tests.**

A Type testing programme has been set up for all new products of
this type.  The content of this has been discussed with and
approved by major Oil Companies and is seen as providing great
confidence in the finished product.

f.    **Field Testing**

In this area the Oil Companies play an important part, and to
date their support has been readily forthcoming and most welcome.
Their taking the chance in installing these new products, albeit
after extensive test programmes, is seen by the manufacturers as
essential to this product development.

**UTILISATION AND VARIANTS**

As with all specialised products of a high-technology, low volume
nature, the industry demands a multitude of variants to exactly
suit particular requirement (whilst maintaining low cost and good
availability of course !).  It is therefore normal for a
manufacturer to develop a basic principle and to provide a number
of detail variations to the main theme, mainly in terms of
mechanical interfaces.

This does provide the manufacturers with a number of problems in
that each major requirement must be treated as a project, rather
than simply providing a quantity of items from a catalogue.  On
the other hand it does ensure that the manufacturers are
constantly on the look-out for improvements in design and
materials.

As mentioned previously, the trend does now appear to be that of
greater confidence in the connector design allowing its use in
more vital areas.  This in itself has increased the number of
variants on the basic theme by virtue of differing
Power/Environment/Deployment conditions.

The connectors are now in use for Umbilical Terminations,
Downhole Instrumentation, Valve Position Sensing, Power
Transmission purpose amongst many others and their use is
actively being considered for Downhole Pumps and Booster Pumps
again amongst others.

Installation of the connectors subsea can take a number of forms,
the most common being :

a.    Diver Installed
b.    ROV Installed
c.    Stab-Plate Installed
      (See Figures 3-7)

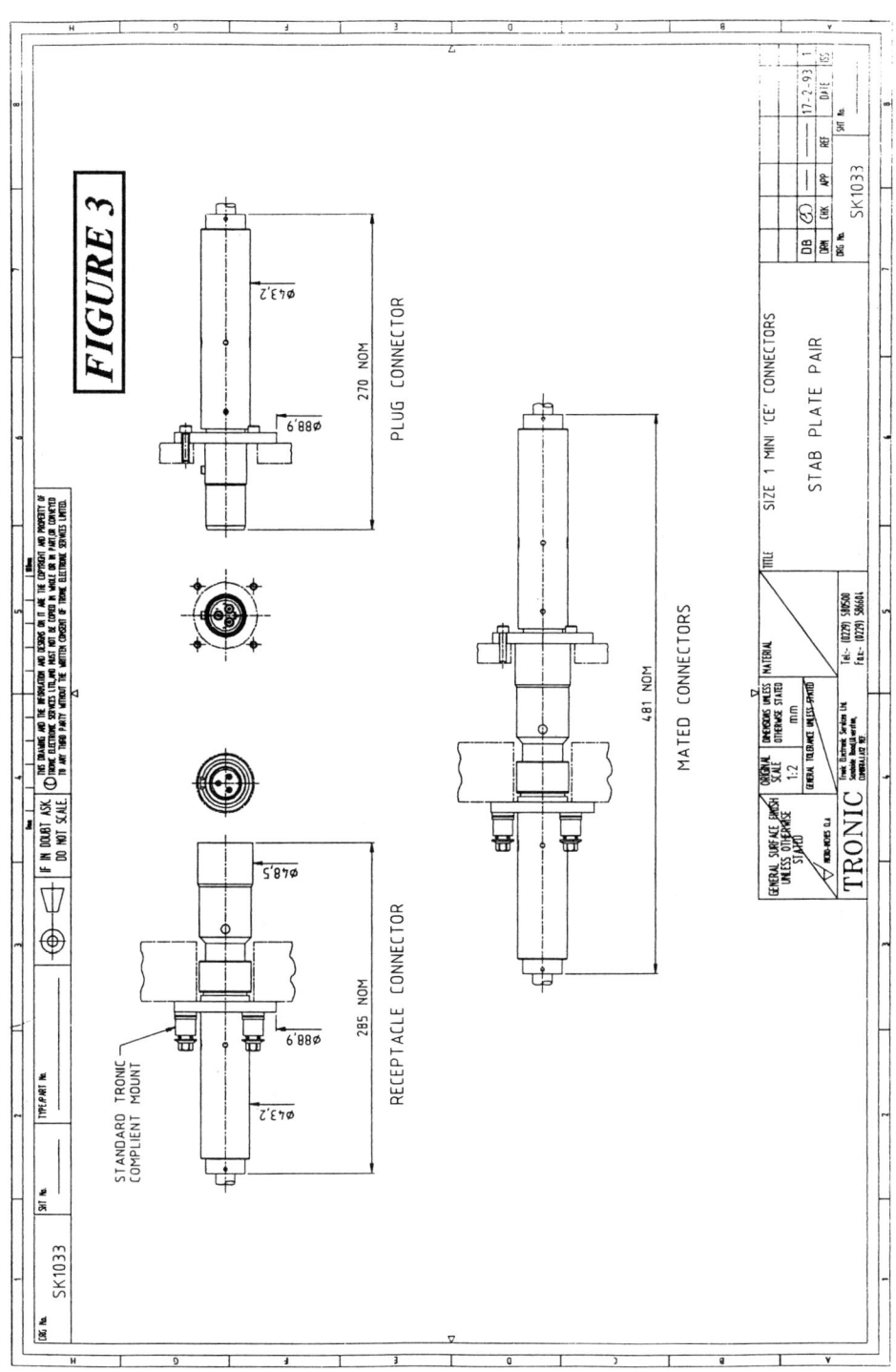

FIGURE 3

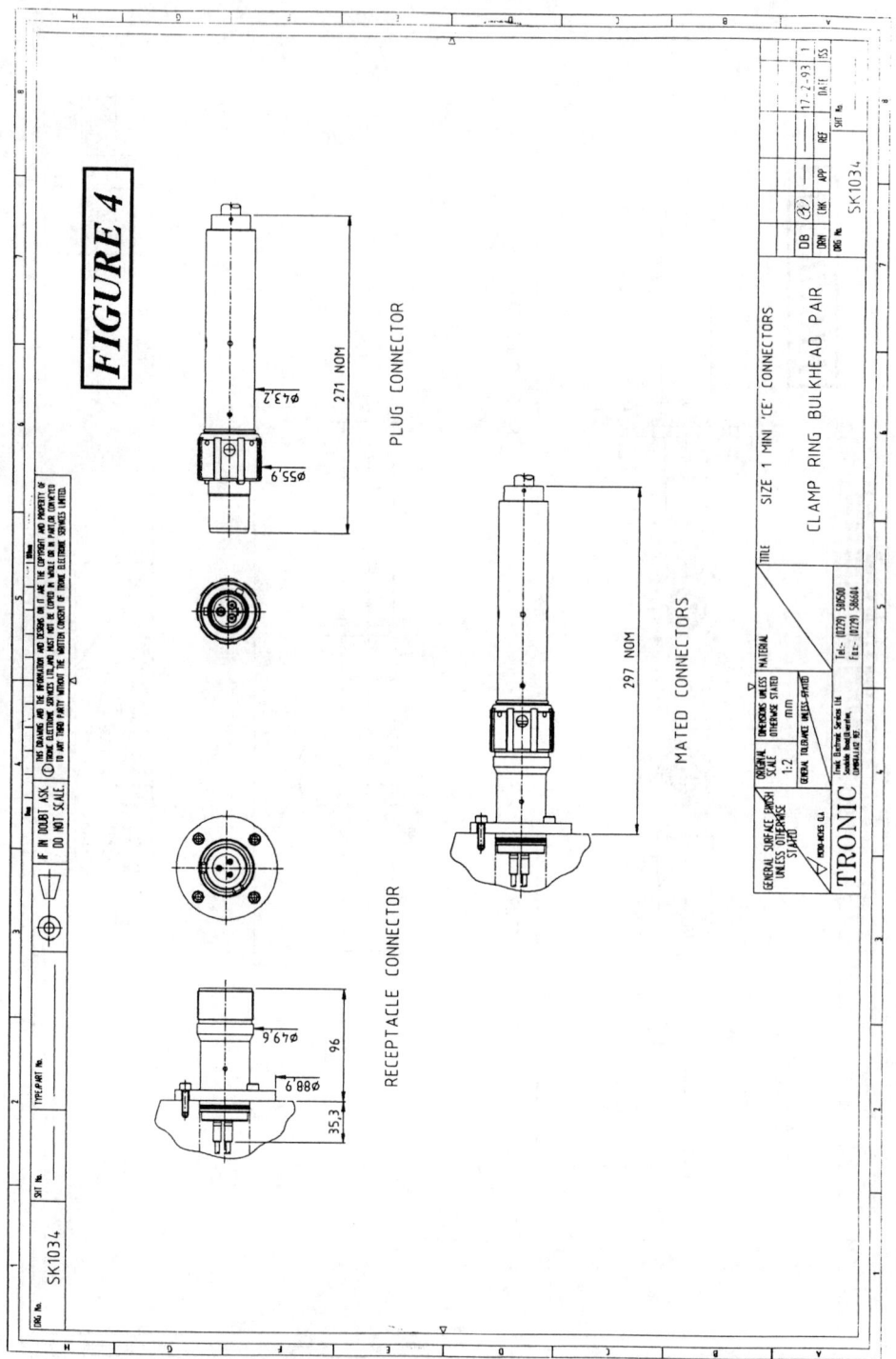

FIGURE 4

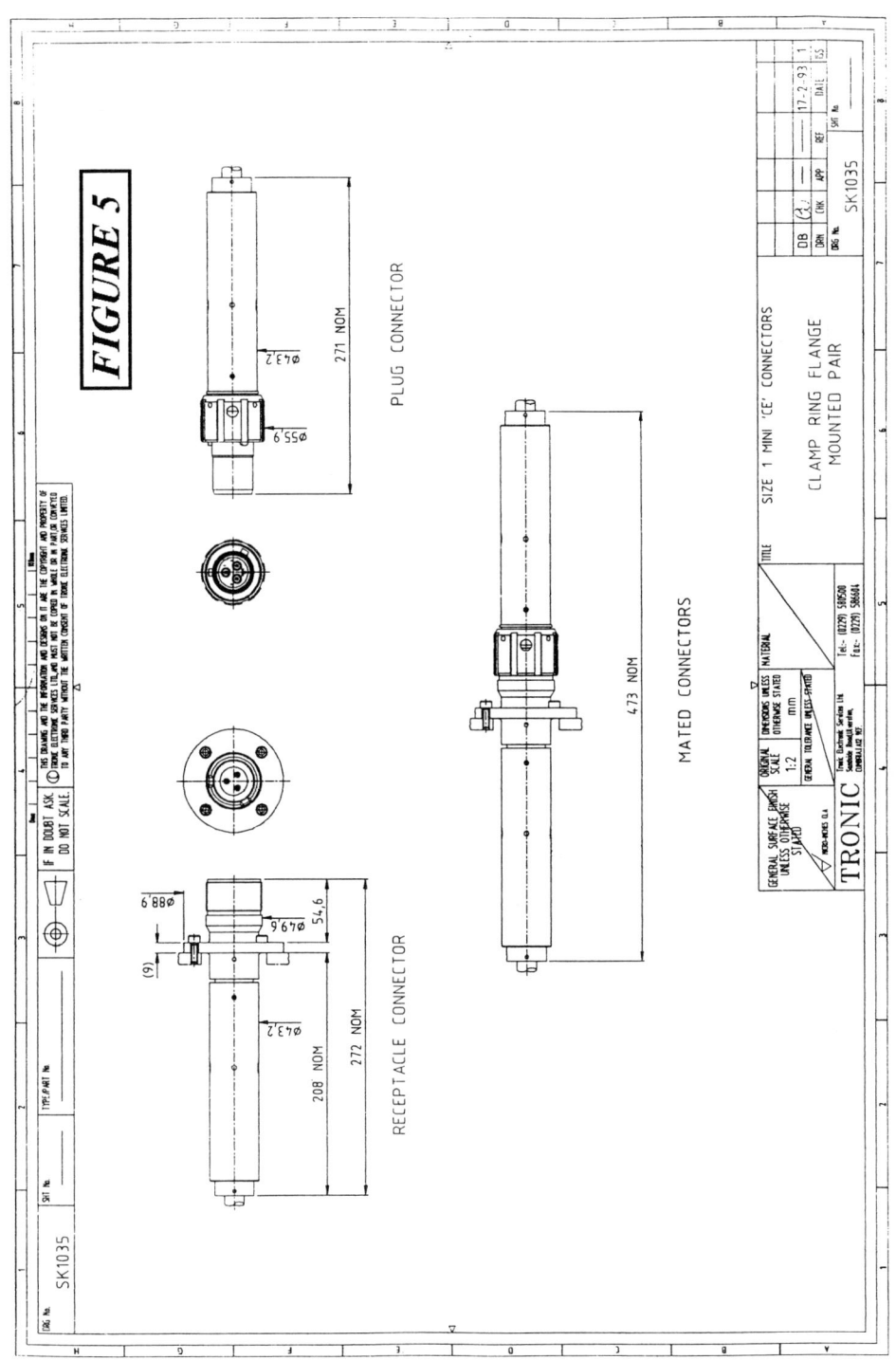

FIGURE 5

PLUG CONNECTOR

RECEPTACLE CONNECTOR

MATED CONNECTORS

SIZE 1 MINI 'CE' CONNECTORS

CLAMP RING FLANGE
MOUNTED PAIR

TRONIC

SK1035

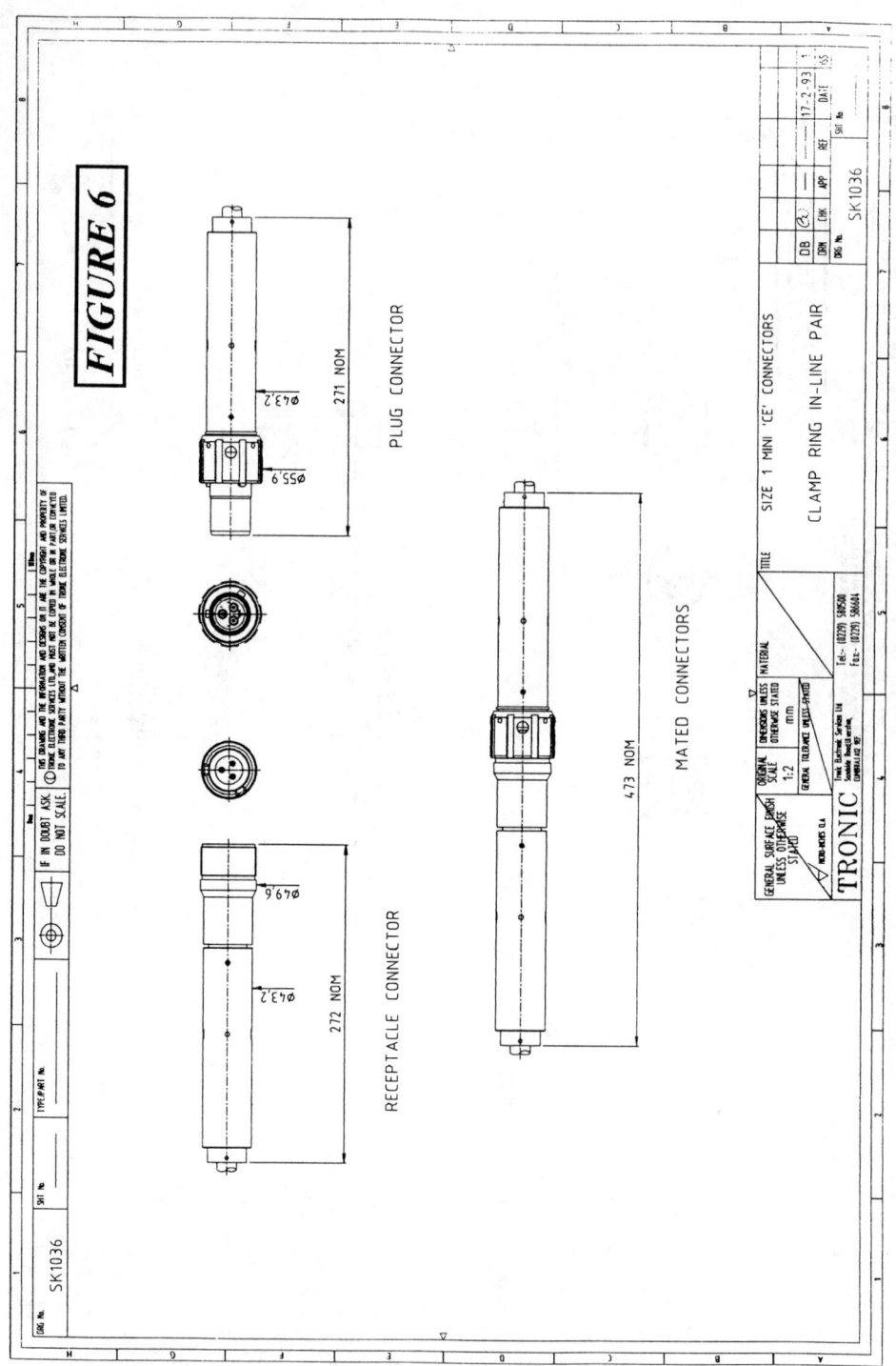

FIGURE 6

PLUG CONNECTOR

RECEPTACLE CONNECTOR

MATED CONNECTORS

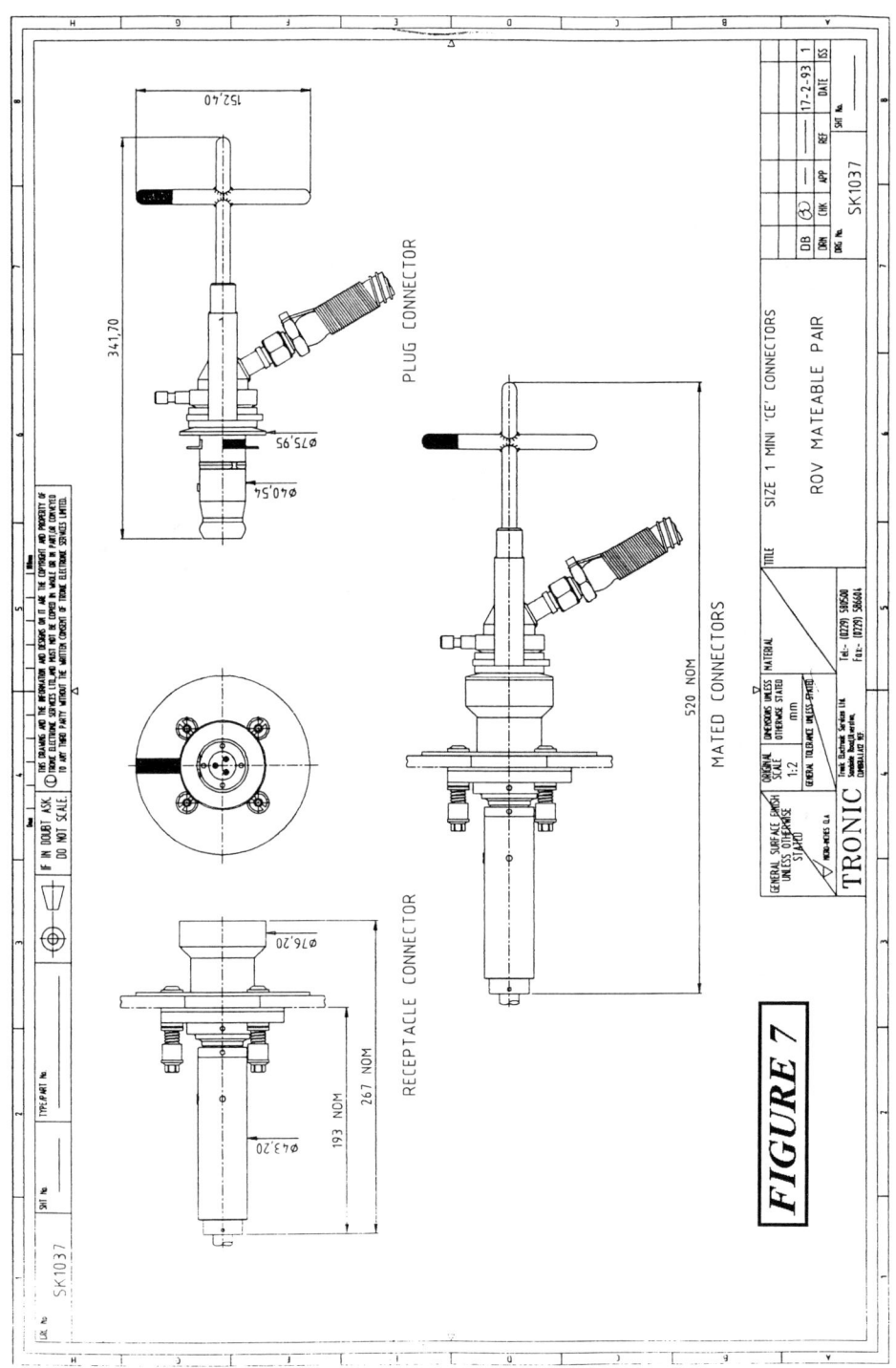

FIGURE 7

All of these have given rise to sub-variants of the basic
connector which are discussed below:

a.   The Diver Installed variant is in some ways the simplest.
     In terms of tolerance to misalignment it is not the most
     demanding, but the strength limitations, access restrictions
     and, of course, physical endurance of the diver must be
     considered.  Mating forces are an important consideration.

b.   Installation by ROV demands great attention to the
     robustness and handlability of the connector.  It must
     have easy and ultimately accurate alignment of keyways
     and these must be clearly visible to the ROV pilot at all
     times.  The interface between the connector and handling
     tool (normally a hydraulic manipulator) is of primary
     importance and is at present the subject of constant
     discussion and revision.  A degree of flexibility is
     essential in one or both halves of the connector,
     normally to be provided by sprung, compliant mountings
     on the receptacle.

c.   Stab-plate installation is gaining rapidly in acceptance as
     a method of quick, accurate and simultaneous connection of
     both Hydraulic and Electrical circuits.  Providing accurate
     guidance and a measure of compliance is incorporated this
     method requires the minimum of skill from operators.  It
     does however require special deployment and retrieval
     tooling and has to be incorporated as part of the design
     at an early stage.

     Stab-plates allow for easy retrieval of modules for repair
     or maintenance, possibly reducing their reliability
     requirements a little.

     The Fluid-filled connector has greatly improved the
     installation of systems by the above methods by removing the
     need for high mating forces and by providing consistent
     electrical properties.

## REPLACEMENT OF EXISTING TECHNOLOGY

In order to exploit the remaining, smaller resources of the North
Sea (and other areas of the World of course) technology has had
to advance at a rapid pace.  Many diverse developments have taken
place to provide the necessary capabilities and more are still
under consideration.

Prior to the availability of the Fluid-filled connector much use
was made of the Inductive coupler as a means of cable
termination.
This device provided the industry with its first, really reliable
coupling and it became very widely used.

However, the Inductive Coupler suffers from a number of
significant drawbacks :

a.     The efficiency of the coupler is in the order of 70%.
       This is not too much of a problem in the case of a
       single joint, but if more than one is required in
       series the losses in the coupler soon become a very
       significant factor.  These result in the need for
       much larger cross-sectional areas in the Umbilical
       cables, a factor which is of enormous significance
       with the greatly increased lengths required in today's
       Satellite fields.

b.     Remote handling is now an established method in the
       Hook-up of Subsea facilities.  Here, the size and
       weight of the inductive device is of significance.
       The fluid-filled connector provides a much smaller
       item to handle, although it has been found from
       experience that size cannot be reduced beyond a
       certain point without introducing connection
       difficulties.  Robustness must not be compromised,
       this being a feature with which the Inductive Coupler
       is well endowed.

c.     The number of circuits and their current-carrying
       capacity is severely limited in the inductive
       coupler whilst multi-pin conductive variants are
       now in service.

d.     A factor never to be forgotten is of course, cost,
       and again the conductive connector has an enormous
       advantage here.  With some recent offshore developments
       requiring large number of these devices the savings to
       be made are enormous.  However, these savings cannot be
       seen to be made at the expense of reliability - hence
       the extensive testing and understandable initial
       caution shown by the industry before its acceptance of
       a product.

It must be remembered though that the development of the Inductive Coupler allowed terminations to be carried out with confidence at a time when no practical alternative was on the market. Its use will most likely continue for some time yet in the most critical applications, and with the more cautious users.

Developments to the "second generation" connectors, the pin and socket metal shell type, allowed their use on subsea applications in the early 1980's. However they were only used on non-critical applications such as valve position sensing and general instrumentation. The power applications were still fulfilled by Inductive Couplers due to their inherent reliability.

It was the relative un-reliability of the earlier of these connector types and their un-suitability for remote handling which spurred the Oil Companies in to their sponsorship of the development of a replacement.

## CONCLUSIONS

The key word in the minds of designers of Offshore and, in particular, Subsea facilities is CONFIDENCE. In a number of cases of course, this is linked inescapeably with another, even more important consideration, that of SAFETY.

In the connector field however, the costs of failure are increasing rapidly with the growing reliance on Electrical systems. Whilst the cost of a connector is by no means irrelevant it pales into insignificance when related to the cost of even one days lost production.

The increase in the level of CONFIDENCE which is brought about by the introduction of the connector under discussion is not only due to the use of new materials and the extensive testing carried out, but to the introduction of multiple redundancy into the device.

Each level of protection is to be capable of independently maintaining the electrical properties of the connector and the proving of these has been a major part of the test programme. For example, the 6 year test at Porsgrunn was carried out with the cable sheath deliberately penetrated thus allowing water to reach the internal boot seals.

In-house testing has included completely filling the cable termination with water, thus relying on individual conductor boot seals, and the flooding of one of the two independent oil-filled chambers on the plug nose.

Further improvements have been identified and are being
actively pursued in order to further increase this level of
confidence.

At all times the involvement of the end-user is actively
encouraged, both in terms of general exchange of views and,
of course, the sponsoring of further development and
representative testing !

References:  Burgess, R.F.   -   Ships Beneath the Sea

# Session 3
# Technology Transfer and Development

# DEVELOPMENT TRENDS IN SUBSEA WELL HEAD CONTROL SYSTEMS, THE SAFETY IMPLICATIONS.

EUR. ING. P.T.  GRIFFITHS BSc, C.Eng. MIEE. M. Inst. M&C,
Petroleum Specialist Inspector,  Technology Branch,
Offshore Safety Division - Health and Safety Executive.
Ferguson House, 15, Marylebone Road, London NW1 5JD

## ABSTRACT

The paper seeks to examine and discuss the safety implications of current developments in Subsea Wellhead Control Systems employed both in production and development drilling.

Such developments include the use of power line signalling, subsea power generation with acoustic signalling, and the incorporation of both surface and subsea programmable electronic systems (PES's)  to control and monitor wellhead valves and blowout preventer stacks (BOP's).  Such systems are said by their proponents to offer significant operational and cost advantages over the conventional remotely piloted hydraulic systems, mainly by the elimination of costly umbilicals.

The various safety related and safety critical functions demanded of such systems are defined, and  the ways in which they are implemented in both present and proposed systems examined. There follows  a comparison of their merits from a safety viewpoint, and a discussion of the design principles needed to ensure that risk  has been reduced to a level which is as low as reasonably practicable (ALARP), and that the remaining risks are tolerable.  To meet these criteria, a clear distinction is required between safety critical and safety related functions in the design specification, with simplicity a major requirement for safety critical items.

Reference is made to the  current regulations and guidance, including Offshore Installations (Safety Case) Regulations 1992, and  Programmable Electronic Systems (PES) in Safety related Applications, General Technical Guidelines.

*Volume 32: Subsea Control and Data Acquisition,* 115–133.
© 1994 *Society for Underwater Technology. Printed in the Netherlands.*

The paper concludes that by following the stated design principles and the proper application of quality systems, it is possible to produce equipment that can be shown to be acceptable as part of a Safety Case submitted under the requirements of the Offshore Installations (Safety Case) Regulations 1992.

## INTRODUCTION

There is a steadily growing requirement for the use of subsea well completions in the UKCS. This is driven by the need to develop the smaller fields in a commercially stringent environment, and is particularly appropriate when a new field can be tied back to an existing offshore process facility.

### Production Systems

The main system parameters that require enhancement to achieve commercial viability are :

- ◆ Increased distance from wellheads to production facility.
- ◆ Simplification of Umbilical requirements.
- ◆ Improved information retrieval from Wellheads.
- ◆ Minimise capital and operating costs.
- ◆ Improve system safety levels.

There are a number of methods currently proposed to achieve these objectives, all have advantages and some disadvantage from a commercial viewpoint. This paper will outline them, and compare and contrast their safety implications.

### Development Drilling

Whilst there is not the pressure to provide lateral distance between drilling rigs and wellheads, the increasing water depths in which wells are being drilled has provided a demand for more sophisticated control systems for Blowout Preventer Stacks (BOP's). This has produced designs which include subsea electronic processing power within BOP systems. The safety implications of this development will be considered.

## SYSTEM DESIGNS

### Subsea Production Wellhead Controls

The basic requirements for these systems are to provide reliable operation of the well
shutdown valves and flow control chokes, and to feed back information regarding the
flowing pressure and temperature of the well fluids. They will also have a safety related
function, and possibly a safety critical one, which must be established early in the design
phase.

A common factor in all designs within the authors knowledge is the use of hydraulic
actuators to operate the valves.   Thereafter there are a wide variety of approaches to the
operation of the controls.

### Traditional Approach

Traditionally, control has been exercised by an extension of the systems developed for
BOP's. This consists of a main high pressure hydraulic supply line fed from the surface,
providing fluid to accumulators in the subsea package. This supply is then applied to
individual valve actuators via a set of hydraulically actuated pilot valves activated by
direct lines from the surface. **Fig 1** gives a diagrammatic representation of such a
system.

The limitation of this system lies in the distance restriction for reliable control of the
pilot valves, due largely to the charging effects when pressurising the umbilical, and the
cost of the complex umbilical required for extended distances.

## DESIGN EVOLUTION

### First Stage.

The first stage in the resultant design evolution has been the introduction of electrically
operated fail-safe SOV's in the subsea package, controlled directly by individually
cabled signals from the surface panel. **Fig 2** illustrates a typical system, which may or
may not also include a master S/D sov. The umbilical can also carry well information
back to the platform. The disadvantage with this system is in the cost of a suitable
umbilical for any more than a few kilometres of separation.

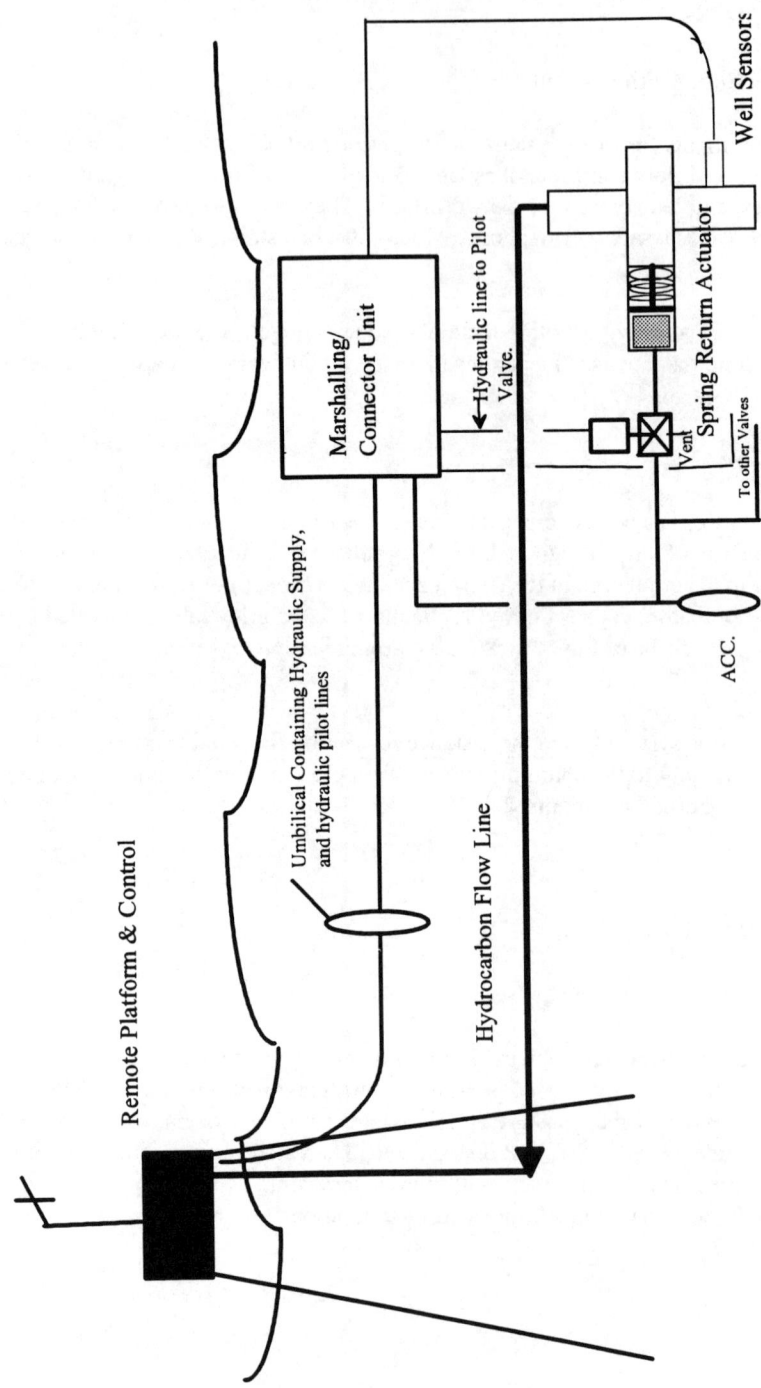

**Fig 1    Diagrammatic Representation of Direct Hydraulically Piloted Subsea Control System.**

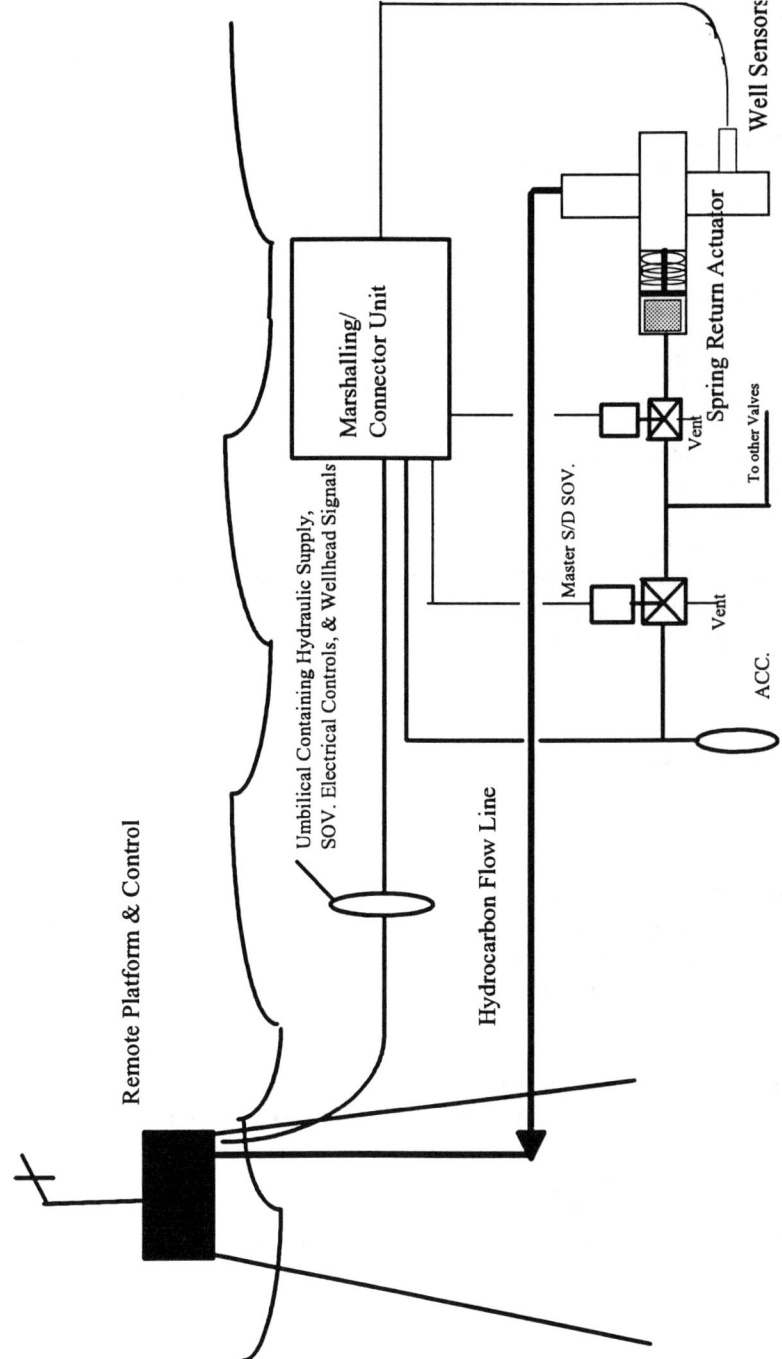

**Fig 2    Diagrammatic Representation of Direct Cabled SOV Controlled Subsea System.**

The umbilical design can be simplified by using a data highway communications link between two processors, and a power line to supply a subsea hydraulics package. **Fig 3** illustrates such a system.

A further simplification of the umbilical can be achieved by superimposing the communications information onto the power supply line, thus reducing the cable and connector requirements to a single pair. With this arrangement a master S/D SOV can be used to ensure S/D if there is a fault in the PES. **Fig 4** illustrates such a system.

### Elimination of Umbilical.

For the longer range applications there is an advantage to being able to eliminate umbilicals entirely. At present there are two methods proposed to achieve this.

**Fig 5** illustrates a system that relies on acoustic coupling for data transfer, utilising power generated subsea.

**Fig 6** illustrates a simple approach, which effectively provides a locally anchored not normally manned installation, in the form of a buoy. This contains all the equipment associated with the wellhead control, relying on a short hydraulic/instrumentation cable to actuate the wellhead valves and chokes. Communication to the controlling platform is by multi-path radio links from the buoy.


## SYSTEM CONSIDERATIONS

### Power Supplies

Having established that electrical power can be used subsea to feed hydraulic packs, a suitable method of supply is required. This can be achieved by cable from the surface, or by generation subsea at the wellhead.
Surface supplied cabling has the cost penalty of a reliable umbilical if the distance is greater than a few kilometres, but may be more suitable for a long life multi wellhead installation.
Local generation can be achieved using either a thermo-electric effect system on production wells, utilising the flowing temperature of the produced fluids, or turbine driven generators inserted into the flow stream of a water injection well. Both of these will also require subsea battery capacity to provide for a 'black start ' of the wellhead. This approach may prove beneficial for isolated individual wells, with a short life expectancy, where the equipment could subsequently be relocated to another site.

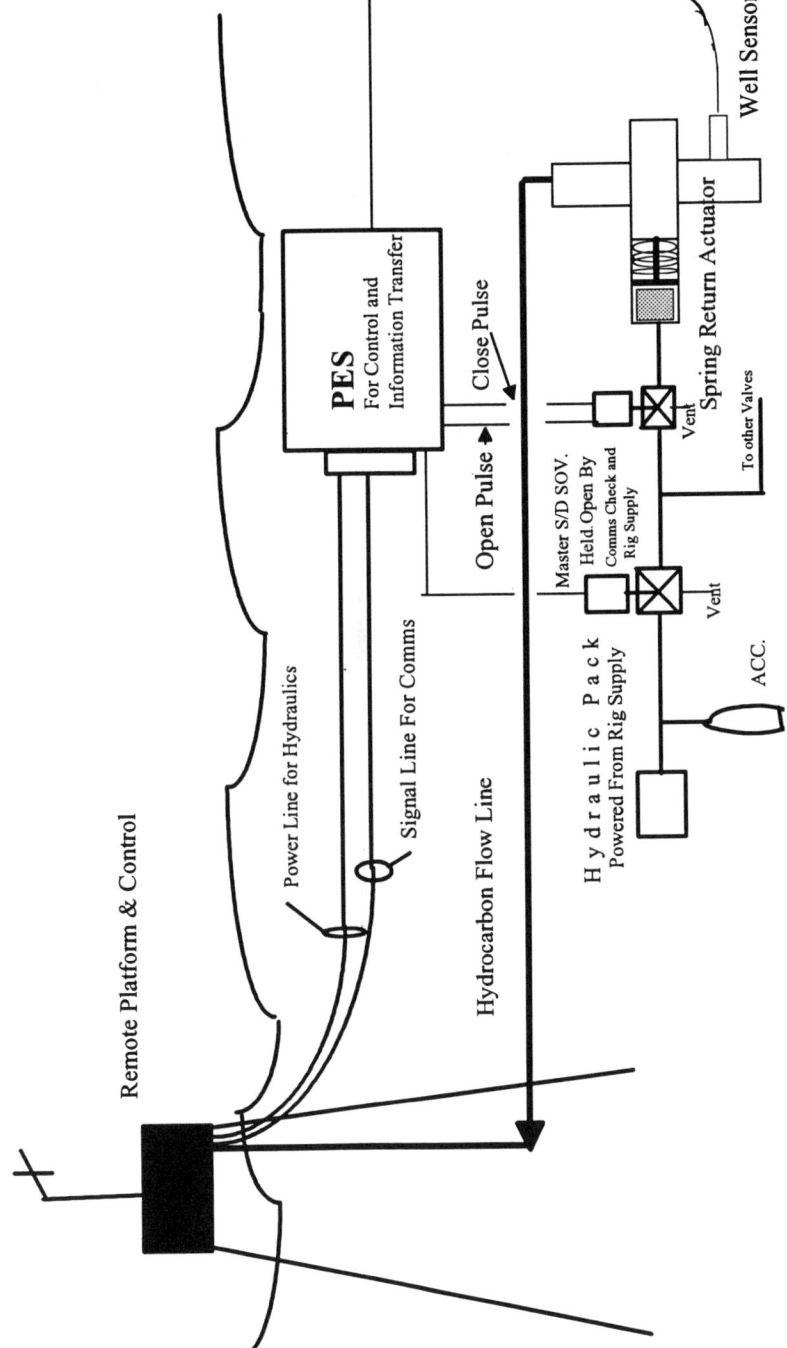

**Fig 3   Diagrammatic Representation of Subsea System, Separate Power and Signal Lines.**

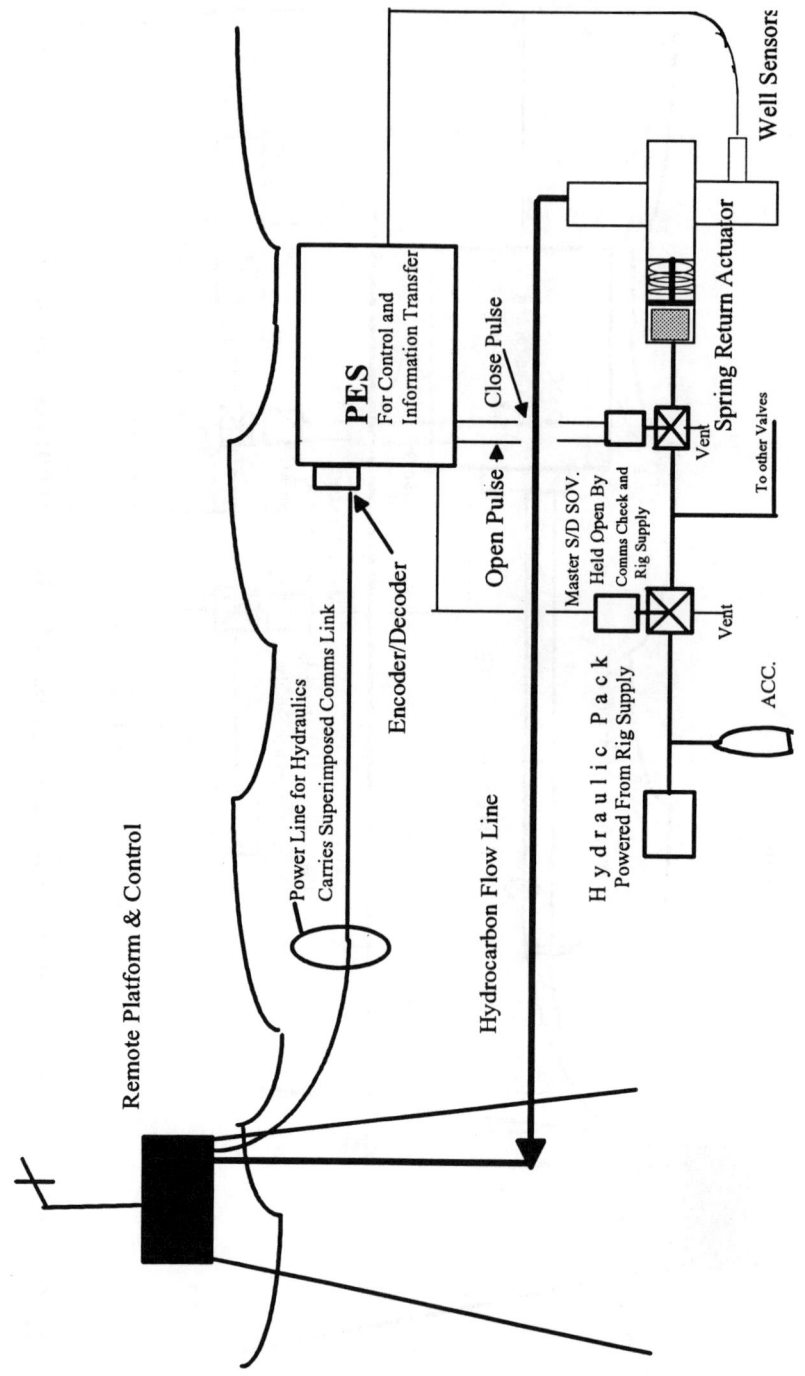

**Fig 4**    **Diagrammatic Representation of Power Line Superimposed Subsea Control System.**

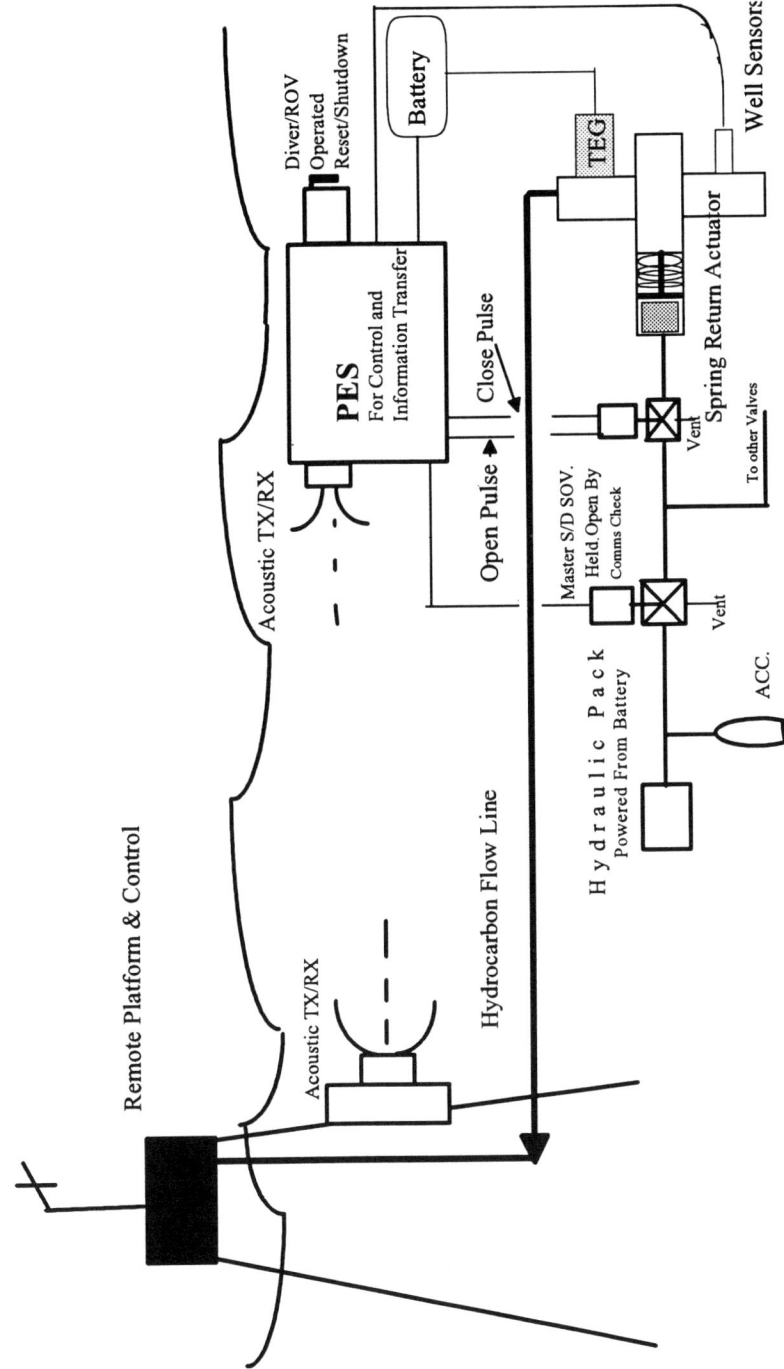

**Fig 5    Diagrammatic Representation of  Acoustic Controlled Subsea System.**

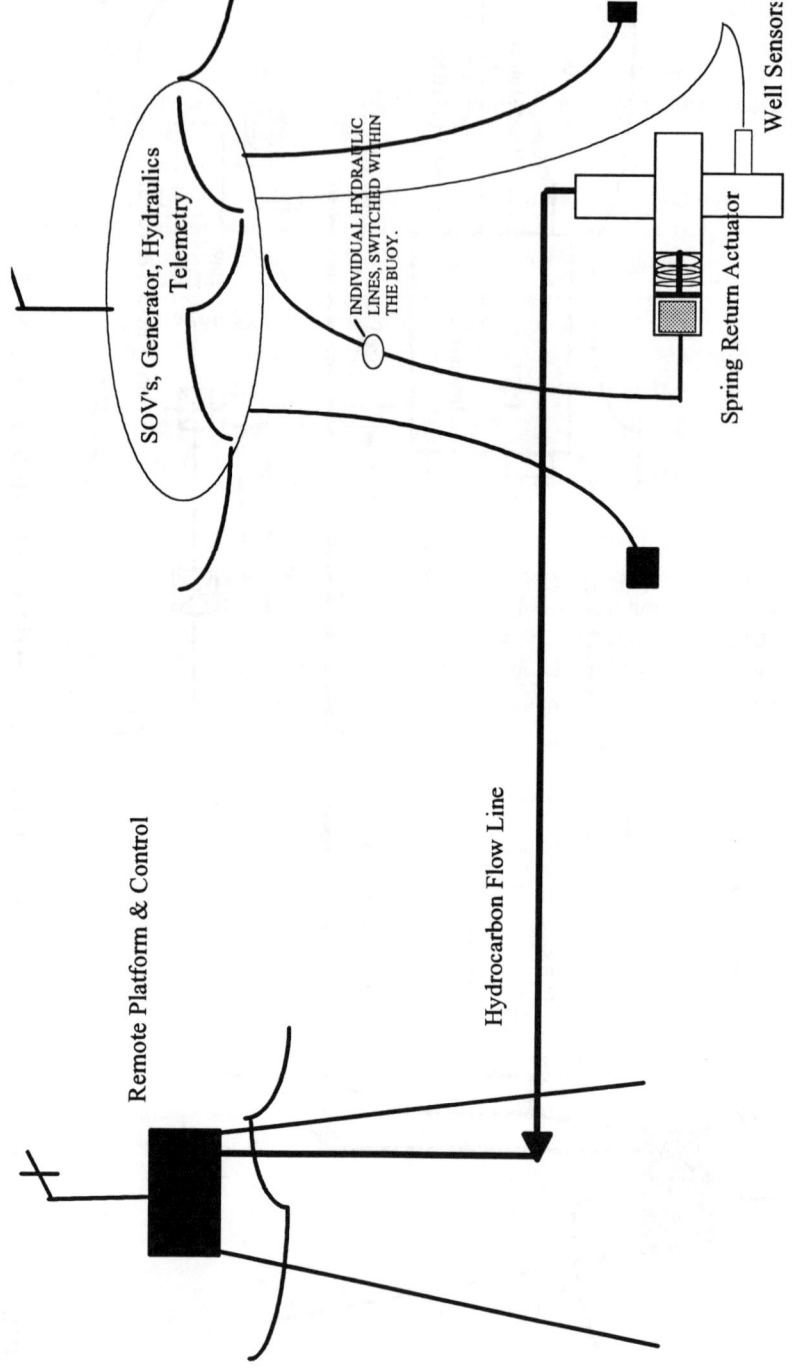

**Fig 6    Diagrammatic Representation of Buoy Controlled Subsea System.**

## Subsea Electronic Processors

All systems providing information gathering/transmittal and control of local solenoid valves via a data communications highway, rely to an extent on the utilisation of electronic processing capacity subsea. They normally also provide subsea 'intelligence' for the control of choke valves, by responding to changes of 'set point' transmitted from the platform. To reduce communications complexities wellhead information is updated only at discreet intervals of say 15 minutes, or either on demand from the control room or on a significant change in a monitored variable. The reliability of these subsea processors is dependent on the standard of their software, as well as their robustness of construction. The use of watch-dog systems is desirable, to monitor the health of the processor. Dependent on the safety criticality of the application, a watchdog system may also be required for safety reasons.

## Communications Methods

There are a number of different methods of communication available between the platform and the subsea processor. These include the use of a data highway in the umbilical, superimposing high frequency communications signals onto the power supply cable, acoustic transmissions from the platform to the wellhead and radio communication to a buoy floating above the well head. All have advantages and disadvantages, but the safety considerations are common to all.

## Subsea Drilling BOP Design

The fundamental difference between wellhead control valves and BOP's, is the requirement for very high pressure hydraulics to close BOP's against the drill pipe, or in extreme cases to shear it, whereas wellhead valves close by spring return on the release of hydraulic pressure. There is a very significant safety risk if the system fails to operate on demand, and a serious cost penalty if it operates spuriously.

Thus the assurance of reliability with BOP's requires a more rigorous engineering solution, resulting in dual umbilicals from the surface control system, and simple hydraulically piloted control valves mounted on the BOP stack. **Fig 7** illustrates a typical design based on this principal. API Recommended Practice 16 E (RP 16E) (Ref. 4) is the established design code for this equipment.

However, for deep water applications (in excess of 1000 ft), this system does not provide adequate closure times for BOP's ( normal API requirement being less than 45

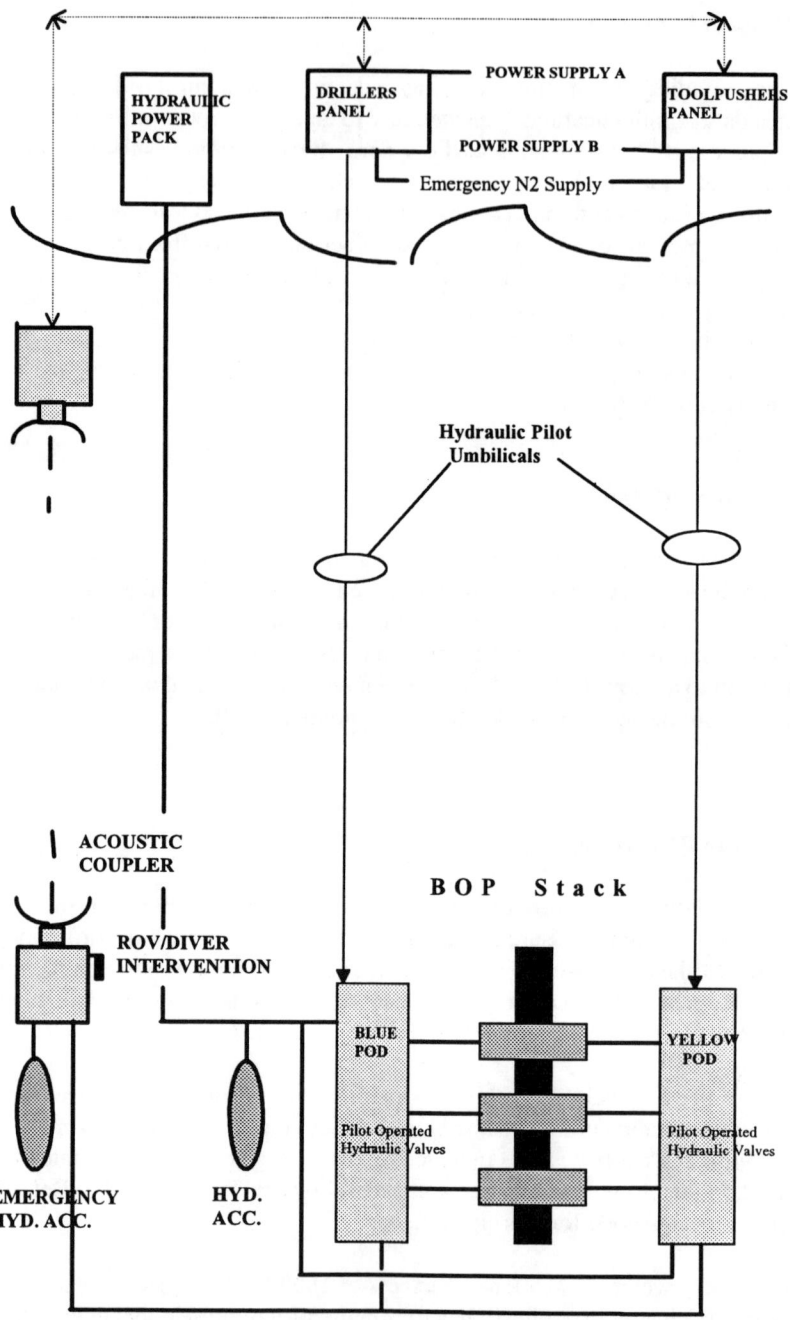

Fig 7 **Representation of Traditional Hydraulicly Piloted BOP Stack**

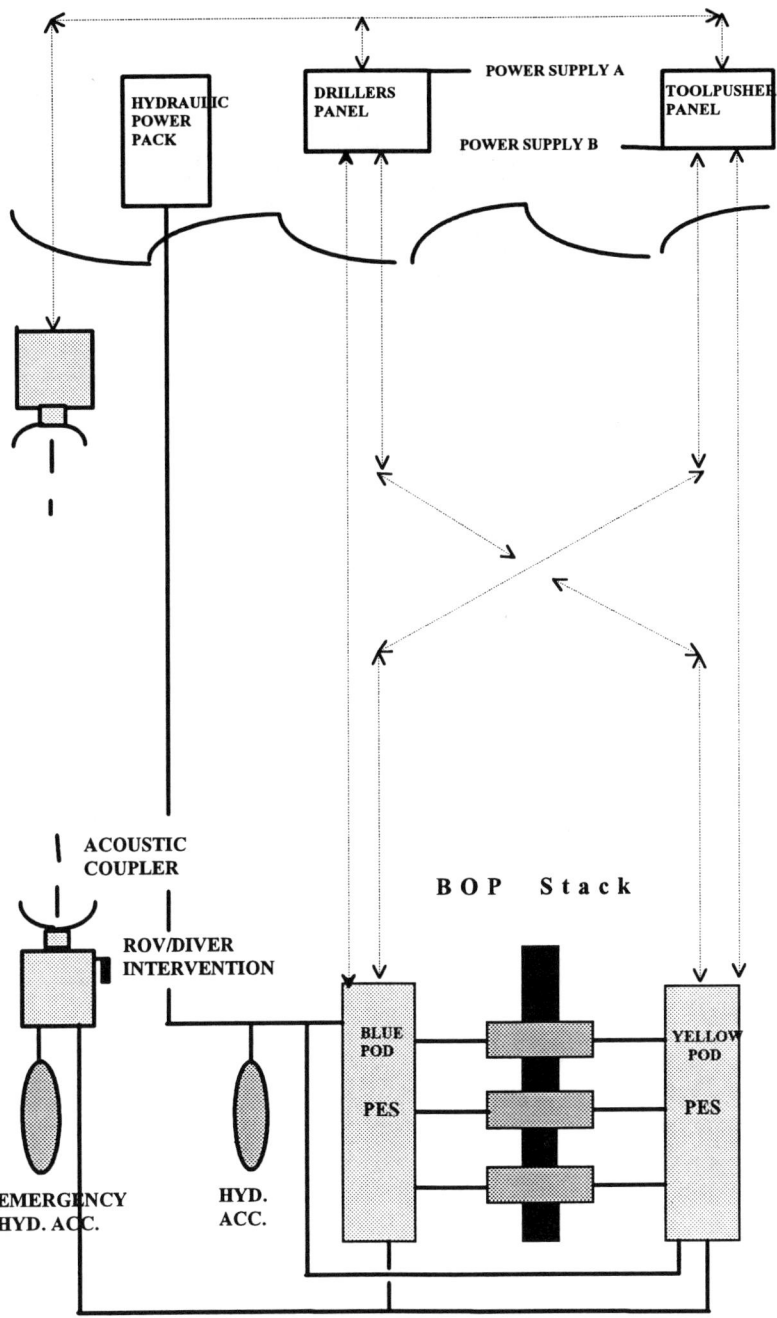

Fig 8 **Diagrammatic Representation of PES Controlled BOP Stack**

## TR Impairement

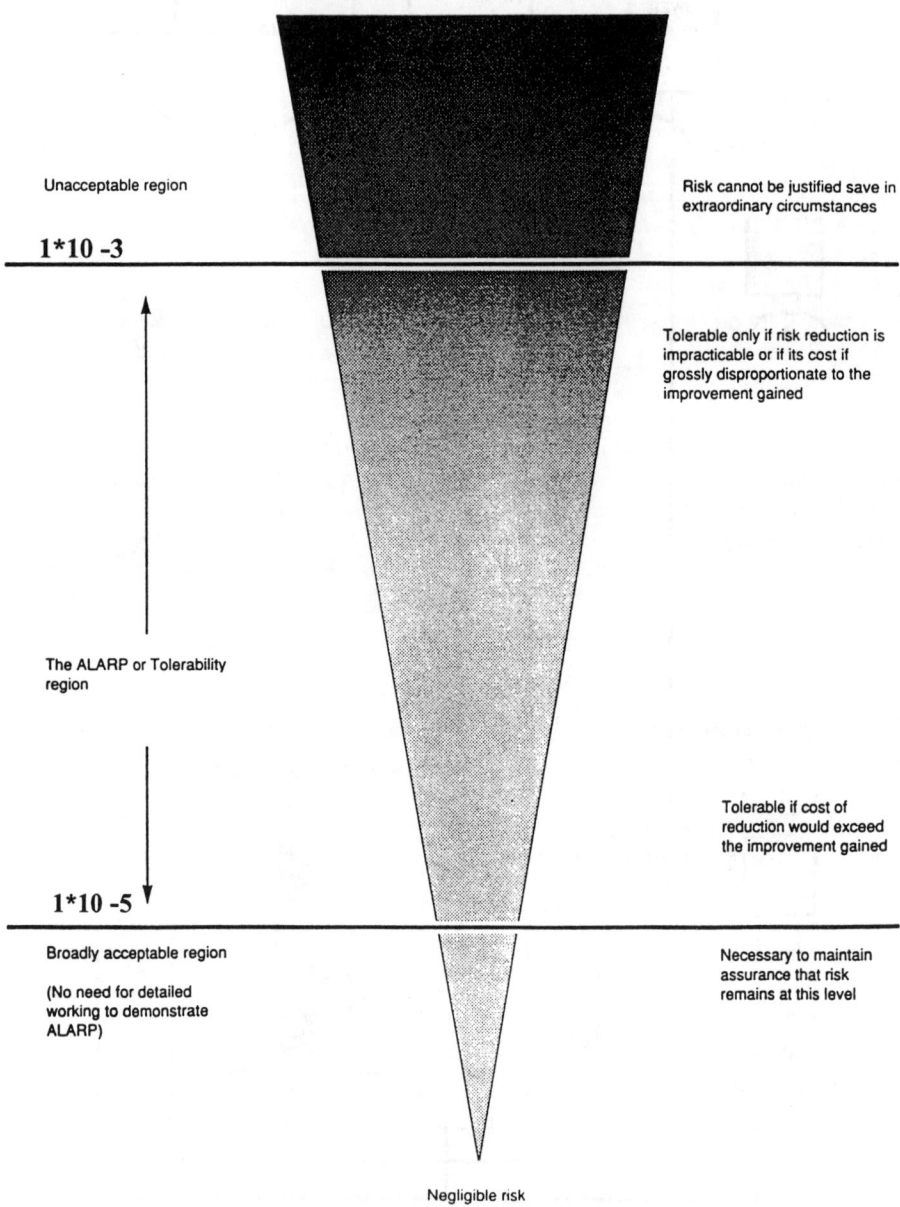

Unacceptable region

Risk cannot be justified save in
extraordinary circumstances

**1*10 -3**

Tolerable only if risk reduction is
impracticable or if its cost if
grossly disproportionate to the
improvement gained

The ALARP or Tolerability
region

Tolerable if cost of
reduction would exceed
the improvement gained

**1*10 -5**

Broadly acceptable region

(No need for detailed
working to demonstrate
ALARP)

Necessary to maintain
assurance that risk
remains at this level

Negligible risk

## Fig 9    The ALARP Triangle

seconds). In these conditions data highway communications and subsea processors are now being applied. This approach will require stringent safety evaluation. **Fig 8** illustrates a possible implementation of the concept.


## DESIGN FOR SAFETY

The requirements of the safety case regulations (SI 2885/1992) are that the risk to personnel following from major accident must be shown to have been reduced to as low as reasonably practicable (ALARP), and certainly to within the tolerability band, as indicated on **fig.9**.

This demonstration is to be achieved by the application of Quantitative Risk Analysis (QRA) (SI2885. Schedule 1. 12) to the accident scenarios identified via analysis of the individual major hazards. The performance of the shutdown system is a significant factor in these calculations. The ALARP principle requires that where it **is reasonably practicable** to improve safety performance by design improvements, this should be continued until the overall risk to personnel (taking Temporary Refuge Impairment as an example) has been reduced to around 1 in 10 *-5 per year. Considerable diligence is needed in the design of control systems to achieve an ALARP solution.


### Safety Criticality

The wellhead control and shutdown systems will have a varying effect on the overall system safety, dependent on their contribution to the field safety strategy. Thus each complete installation design concept must be assessed individually to determine whether the wellhead control equipment is **safety critical** or **safety related** , in terms of its contribution to the total platform risk levels. Appendix 1 extracts the definitions of system categorisation from Ref. 2, and defines safety critical and safety related in terms of the system categories. It is important to realise that it is the requirements of the process that must initially be categorised, and the protection system designed to meet these requirements.

Factors which can influence the categorisation include proximity of wellheads to manned platforms, provision of Riser ESDV's and subsea isolation valves on flow lines and the position of specification breaks between wellhead manifolding and subsea flow lines.

For example, a wellhead with a closed in pressure of 330 bar feeding a sea line designed for a max. operating pressure of 130 bar, with no other means of relief, would be considered as a category 1 requirement for protection. It follows that the control of wellhead shutdown would be a safety critical system, and would need to be designed accordingly.

Because of the potentially dire consequences of a failure to act on demand, BOP systems will usually always be considered safety critical.

Therefore following the specification of functional requirements for a subsea system it is vital that its safety criticality is established.    Following the guidlines set out in ref 2. this can be classed as either Category 1, 2 or 3. See Appendix 1.

Where programmable electronic systems are being used for Category 1 and  high impact category 2 equipment (ie. Safety Critical), it will be necessary to apply the design principles outlined in ref.1, which essentially requires the provision of a diverse method of activating complete shutdown. This diverse method should normally be a fail-safe system.

For all safety critical systems, there must be a calculation to evaluate the potential the equipment failing to act on demand. The following factors should be included when developing the fault tree:

+ Expected Demand Rate.
+ Designed Proof Test Interval.
+ Reliability of Individual System Components. (From a reliability data base)

The result can then be used to determine if the system provides sufficient risk mitigation to reduce the overall risk to a tolerable level. This is achieved by inputting the calculation results into the  QRA calculations.

## System Configurations

It is important to realise the distinction between operational reliability and safety criticality.

When considering system configuration of safety critical systems, the  engineer is concerned with their safety perfomance. If they are unreliable, this is not so important, providing they only fail to a safe state , with a revealed fault.  For this reason, the more complex triple redundant systems that have gained acceptance, are not necessarily the best type of system from a safety standpoint. The very complexity that improves their system reliability can mitigate against their ultimate safety, and makes them less favoured for safety critical applications.

The important requirement for safety critical systems is to include some method of fail-safe diversity into the configuration. Methods of achieving this with subsea systems include, but are not limited to, the following:

1. Permanently energised SOV's to maintain hydraulic pressure on open Spring Return actuators.
2. 'Master' S/D SOV maintained open by fail-safe power supply, where other SOV's require energisation for both OPEN and CLOSE valve actions.
3. Latched open SOV's supplying valve actuators, where the hydraulic pressure can be removed from the platform end of the umbilical to allow closure of all valves.

All of the above will allow a reliable diverse method of closing wellhead valves if the normal communication and control path fails.

There are additional advantages to having fail-safe diversity in a system, even if it is not safety critical. If only one method is available, relying on the communications path, it would be normal to design for the automatic shutdown on failure of that path. When diversity is available it will often be possible to delay such a shutdown for a significant time, sufficient to enable the re-establishment of communication, meanwhile maintaining production. Thus increased safety also assists in increased production stability.

Whilst this approach may appear less safe than going to S/D on communication loss, account must be taken in overall risk analysis that excess spurious upsets to the production system tend to increase the demand rate on other safety systems, with a consequent reduction in overall safety.

## BOP Systems

Because these systems require power closure of the BOP rams, they are denied the simple fail-safe diversity of wellhead systems. Conventional designs to date have relied on duplicated umbilicals to separate control pods within the BOP stack to provide the necessary degree of reliability. When the system is developed to include subsea processing power and data highway communication links, for what is a Category 1 system, the full rigour of the PES guidlines (Ref.1) must be invoked. A system of this type without some diverse method of achieving remote control of shutdown could not achieve a sufficiently low potential of failure to act on demand to satisfy the requirements of the QRA for blowout hazards.

Diversity may take the form of a diverse PES system, as described in Ref. 1, or some other separate system, such as acoustic coupling to an emergency hydraulic accumulator system, as illustrated in Fig. 8.

## QUALITY SYSTEMS

In order to achieve the reliability levels required for safety, as well as reliability, the use of recognised quality assurance techniques (eg Ref 3) is recommended in the design and manufacture of subsea wellhead controls. This applies especially to the design of software, where a formal validation procedure should always be followed.

## SAFETY CASE

The safety case for any Installation or MODU will need to contain details of the design of subsea wellhead or BOP controls (Eg. SI 2885 Schedule 1. 7 and schedule 3. 5), and information on the criticality of the system. Due to the importance that these systems can have for safety, they may be carefully scrutinised during the assessment process to ensure that ALARP has been demonstrated.

## CONCLUSION

Whilst there are a number of diverse system designs available for the control of subsea wellheads and BOP's, some provide a simpler solution to the requirements of safety critical and safety related functions than others. From the safety viewpoint, simple systems with independent diversity are preferred.

When selecting a design type for safety critical or safety related application, it is necessary to consider their relative safety performance, and demonstrate by QRA within the safety case, that the equipment chosen ensures that risks to the health and safety of personnel have been reduced to as low as is reasonably practicable.

Providing the design has been carried out with due diligence in accordance with the recommendations of the reference documents, a satisfactory demonstration of ALARP should be possible.

## REFERENCES

1. HSE: Programmable Electronic Systems in Safety Related Applications, Part 1 An Introductory Guide, 1987. Part 2 General Technical Guidelines, 1987. ISBN 00 11 883906 3.
2. EEMUA Publication No. 160. Safety Related Instrument Systems for the Process Industries (including Programmable Electronic Systems) 1989.
3. ISO 9000-3 Quality management and Quality Assurance Standards - Part 3: Guidlines for the application of ISO 9001 to the development, supply and maintenance of software.

4. API Recommended Practice 16E (RP 16E), First Edition, October 1, 1990.
   Recommended Practice for Design of Control Systems for Drilling Well Control
   Equipment.

## APPENDIX 1

### Extract from Reference 2

**Category 0 System.**  *A self acting mechanical device or system which is a protection against dangers to personnel. Examples include relief valves, bursting discs and containment.*

**Category 1 System.**  *A non self acting system which requires an outside source of energy and which is a protection against dangers to personnel. Such a system may be necessary where self acting mechanical systems are not used or are not adequate when acting alone. Examples include electronic, hydraulic, pneumatic and relay systems.*

**Category 2 System.** *A system which protects against the damage to the environment, damage to the process plant, loss of product or production.*

**Category 3 System.** *A control system which ensures reliable production and maintains plant operation within operational limits.*

For the purposes of this paper it is recognised that Category 2 covers a very wide range of applications, with a consequently wide range of environmental damage and production loss. It is considered that applications which would have a serious impact should be dealt with in a similar manner to those in Category 1. These systems have been designated as **Safety Critical** in common with those in Category 1.

# UMBILICAL-LESS INTEGRATED CONTROL SYSTEM

P. Yates
John Brown Engineers & Constructors Ltd.
Subsea Group
20 Eastbourne Terrace
London W2 6LE

D.M. Jones
Thorn Security
1 Forest Road
Feltham
Middlesex TW13 7HE

## ABSTRACT

Fields currently being discovered in the North Sea are not as large as those of the 1970's and 1980's; they cannot therefore, economically sustain the implementation of a fully functioning platform. The approach currently employed to extract reserves from such 'marginal' fields is to install a subsea wellhead, controlling this via a control umbilical tied back to an existing platform. Although this is less costly than building a platform the cost is still significant. In addition, there are also technical risks involved in laying and trenching the umbilical, which may limit the distance between the remote wellhead and the host platform.

Most, if not all subsea installations require hydraulic or electrical controls. Communications to subsea trees, manifolds or templates involve the purchase and installation of umbilicals. With the average cost of umbilicals alone running at around £120 per metre to purchase and installation costs at least double this figure, depending on lengths and site conditions, it is not difficult to realise costs, when including sophisticated control modules and terminations, of between 5 to 7 million pounds for a multiplexed controls and umbilical system some 20km from a controls source.

This paper highlights the fact that for certain situations, savings in the order of 50% can be made by the acquisition and installation of an Integrated Control Buoy (ICB), together with a more reliable system.

*Volume 32: Subsea Control and Data Acquisition*, 135–153.
© 1994 *Society for Underwater Technology. Printed in the Netherlands.*

# WHAT IS IT?

- **Name : Integrated Control Buoy (ICB)**

- **Subsea Well Control - Umbilical less**

- **Buoy**

    - **Moored in field**

    - **Control, monitoring, chemical injection**

    - **Communication**

        - **Radio, line of sight**

        - **Satellite**

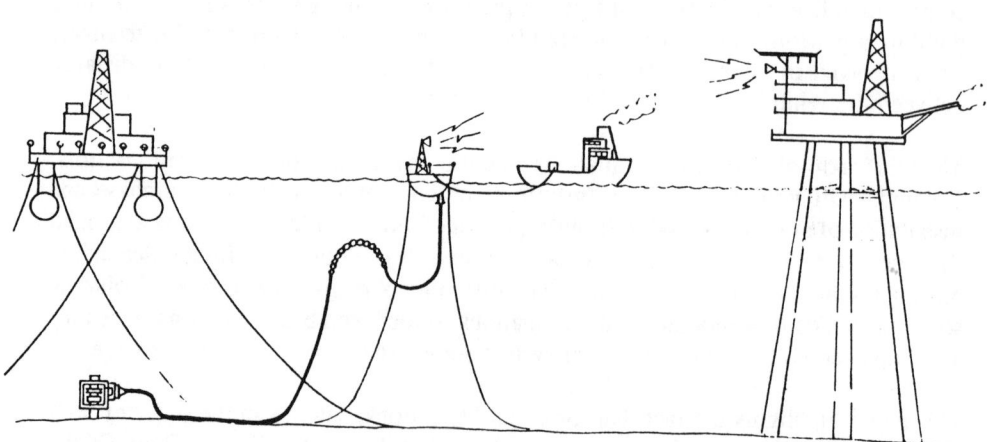

## 1.0  INTRODUCTION

Control of the wellhead via long hydraulic umbilical can be problematical due to the delay associated with the pressurisation of the umbilical. These problems can be reduced, to a degree, by the use of subsea electronics, but the repair of these systems, should they fail, is costly due to the difficulty in accessing the equipment on the seabed.

A solution to this problem is to install all necessary control and monitoring equipment in a relatively small buoy (6 metres diameter and 6 metes high) moored adjacent to the subsea wellhead. This is connected to the christmas tree via a short, vertical, hydraulic and signal riser. The control systems on the buoy are then supervised, via a radio link, from a nearby master control station situated on board a Platform or Floating production facility.

The following technologies are employed in the system:

- **Buoy technology** (including moorings) is well understood through experience gained with the Oceanographic and Meteorological Data Gathering Buoys which were installed and managed under contract to UKOOA.

- **Control and Monitoring Equipment** will be standard, 'off the shelf' equipment as used offshore for many years.

- **Dynamic Riser** technology is well developed in the industry. Recognised suppliers will be used.

- **Radio links** have been supplied for use in many harsh environments, both offshore and onshore, for control and monitoring applications.

## 2.0  BACKGROUND

John Brown presented a paper on the possible use of control buoys in September 1989 at the *SPC offshore Europe conference but the concept was not developed due to lack of historical data on buoy reliability at that time.

Recessionary times and present day economics have revitalised the need to pursue and investigate new and historical innovative economical but reliable ideas.

In line with modern day concepts of "CRINE" (Cost Reduction Initiative for the New Era), Thorn EMI and John Brown are extending the use of the existing

"Oceanmaster" data gathering buoy to become a fully integrated control buoy (ICB) capable of the autonomous control of subsea wellheads. The ICB will interface by monitoring and communicating between the subsea installation and the host facilities.

At the end of 1992 Thorn EMI discussed with JB the use of their "Oceanmaster" data acquisition buoys for the possible use as a subsea integrated control system. A confidentiality agreement was signed between Thorn EMI and John Brown enabling the concept to be developed further.

## 3.0    BUOY TO MASTER CONTROL STATION COMMUNICATIONS SYSTEM

The key to the success of this system is the control philosophy which underlies the operation of the buoy.

It is clear that, because of the height limitation on the buoy mounted antenna coupled with the dynamic sea environment, communications between the buoy and the host platform may not be possible 100% of the time. With the system proposed, routine communication is not necessary 100% of the time.

The buoy's control system is designed to be fully autonomous. That is, it will operate the buoy at the remote well location, without requiring a continuous communication link to the host platform. Safety back-up will be provided by on board Emergency Shutdown and Fire and Gas Systems. All well parameters will be measured and analyzed by the buoy control system and any necessary safety related action will be taken, autonomously, by the buoy. In this way it is not necessary to have fully continuous communication between the buoy and the host platform. Should a hazard befall the platform or its communication system the buoys control system, having been previously programmed, will have control of the wellhead by activating a fail safe situation or shutdown.

Routine communication between the buoy and the platform will be via a Line of Sight microwave link. This will take place at regular intervals, controlled by the buoy and acknowledged by the platform system. The information transmitted will consist of well status (temperatures, pressures, choke valve position) and on board buoy system status (fuel levels, battery voltage levels etc). This data will be fed to the platform Master Control Station (MCS) for display and storage, process data being transmitted into the platform DCS system. However, any exceptional event which occurs will always be reported immediately. If, during routine communication, the platform fails to acknowledge transmission from the buoy, the buoy will repeat the transmission several times in rapid succession. If an acknowledgement is still not received the buoy will place the christmas tree in a Process Hold and await further instructions from the pre-programmed on board safety systems or from the platform.
**\*H.L. Young 1989 "Development In Subsea Controls"**

A back-up radio system based on meteor burst technology will be employed to ensure that emergency communication is available 100% of the time. For practical reasons, this will be capable of limited communications between the buoy and platform (no general system status information will be sent), but will be sufficient to maintain production if the main link fails.

If communications are still not established after a further period the wellhead will be shut down by the on board buoy Emergency Shutdown System by closing the Down Hole Safety Valve and the production and annulus master and wing valves as programmed.

## 4.    INTERFACING WITH HOST ESD & DCS SYSTEMS

Operation of the wellhead will be monitored at the host platform, using the existing platform Distributed Control System (DCS) via the MCS. In normal circumstances, the operator will not be required to initiate specific valve movements, but will instead instruct the buoy to carry out a series of high level functions, for example, start-up, increase/decrease oil/gas output, shutdown, etc. The series of detailed operations required to achieve these high level functions will be implemented autonomously by the buoy's control system.

The MCS will display, print out, alarm and interface all the various functions with topsides controls systems (DCS) including emergency shutdown requirements.

The subsea master controls station (MCS) will provide the facilities for the control and monitoring of the subsea production system (consisting of production wells and/or water injection wells) plus the control equipment for subsea pipeline isolation valves (SSIVs) including oil and gas export.

The base case will be a production or water injection well and 8 subsea isolation valves.

The MCS will also provide facilities for interfacing with topsides DCS and ESD systems.

The MCS will function independently of the platform distributed control system (DCS) and provide all control and monitoring functions of subsea and platform equipment which shall be autonomous from the DCS. Each processor includes a disc unit for file and system storage, programme reloading, and means for archiving data and programmes.

The dual microprocessors and discs are configured to allow "hot standby" operation. It is envisaged that each microprocessor and disc will operate as a unit, the standby machine being kept up to date by continuous processor to processor/disc transmissions. An automatic switch-over to the standby machine will occur in the event of a failure/malfunction of the on-line machine being detected. There will be no preference as to which of the two machine units is on-line and the status of each shall be indicated. Each processor will have a dedicated power supply unit fed from the platform UPS.

The control and information passing across the MCS interface will include, as a minimum, the following classes of data:

- Control and Monitoring;  DCS to MCS

- Operation of Individual Xmas Tree Valves (when on manual override)

- Operation of Individual Manifold Valves

- Operation of Individual SSIVs

- Operation of Xmas Trees in a preselected sequence

- Manual Initiate Shutdown Individual Xmas Tress (normal mode of operation)

- Manual Initiate Shutdown Individual Cluster and Manifolds

- Manual Initiate Shutdown of Individual SSIVs

The MCS interface with the Host facilities ESD System will be by hard wire instrument links and/or RS 232 data links.  The software interface between ESD System and Subsea MCS would be carried out by the host facilities, the interface protocol normally being the industry standard "MODBUS" RTU.

## 5.    BUOY MOUNTED AUTONOMOUS CONTROL SYSTEM

The Data Acquisition and Communications System will perform autonomous control of all valves, and support systems, including emergency shutdown facilities.  In normal operation the status of valves will be changed according to a predetermined set of instructions within the buoys control system.  As the operation of individual valves will not normally be required a manual override facility will be available at the MCS for individual operation of valves if required.  The buoy will have control of the following tree functions.

- Operation of Individual Xmas Tree Valves
- Operation of Individual Manifold Valves
- Operation of Individual SSIVs
- Operation of Subsea Chokes
- Operation of Xmas Trees in an preselected sequence
- Emergency Shutdown Individual Xmas Trees
- Emergency Shutdown Xmas Trees in preselected sequence
- Emergency Shutdown Clusters and Manifolds
- Emergency Shutdown Individual SSIVs*
- Manual Initiate Shutdown all Production Xmas Trees
- Manual Initial Shutdown all Water Injection Xmas Trees and Manifolds
- Choke stepping open/closed to an absolute position

The buoy's data acquisition system will receive data from and send commands to the subsea equipment via the dynamic riser in the form of digital and analogue signals. The buoy's data acquisition system will monitor these signals and take any appropriate action at regular intervals. This information will be sent back to the MCS for display, updating and archiving. For the purposes of the prototype it has been assumed that the ICB will transmit data to and receive data/instructions from a MCS situated on a platform/floating production facility or land base up to 25km from the subsea installation. Data will also be collected from within the buoy (e.g. smoke, collision sensors and environmental housekeeping control data) for regular transmission to the MCS.

The Data Acquisition and Communications System will also provide facilities to monitor the following subsea process parameters.

- Wellhead Flowing Pressure
- Wellhead Flowing Temperature
- Annulus Pressure
- Downhole Pressure and Temperature (1)
- Downhole Pressure and Temperature (2)
- Valve Position
- Choke Position

The analogue and digital data acquisition subsystem will provide sufficient channels for measuring ICB housekeeping functions, plus all remote pressure sensors, remote temperature sensors, valve position sensors, choke position sensors and differential pressure sensors as a detailed below. In addition, all hydraulic, chemical and electrical supplies will be maintained automatically by the ICB.

*NB: This would only be true where a remote SSIV is being controlled via a Buoy.

The following are the analogue inputs that will be incorporated into the Integrated Control Buoy.

| | |
|---|---|
| Downhole Pressure Tans | 0-1000 Bar |
| Wellhead Pressure Transmitter | 0-1000 Bar |
| Downhole Temperature | 0-200°C |
| Wellhead Temperature | 0-200°C |
| Press. Differential Transmitter | 0-690 Bar |
| Press Indicating controller | 0-390 Bar |
| Downstream Press. Transmitter | 0-390 Bar |

All pressure transmitters shall have adjustable high pressure alarm and high pressure alarm/trip set points.

The digital data acquisition system shall provide sufficient channels for controlling and displaying the status of valves, chokes and housekeeping functions.

The following housekeeping parameters will be measured within the buoy and acquired by the data acquisition unit for transmission into the communications system.

- HPU hydraulic supply HP

- HPU hydraulic supply HP (standby)

- HPU hydraulic supply LP

- HPU hydraulic supply LP standby and pump

- HPU reservoir level low alarm and shutdown

- Filter blockage (DP) alarm

- Pump (Hyd) fail alarm

- Pump run

- Generator run

- Generator fail alarm

- Battery Voltage low alarm

- Battery charger fail alarm

- Access alarm

- Collision indicator

- Smoke Alarm

- Generator Fuel Level

- Chemical Storage Levels

## 6.0 HYDRAULIC CONTROLS

A major part of the space available onboard the ICB will be taken up by the hydraulic control system. Subsea valves are actuated by high pressure hydraulics either directly controlled or remotely controlled using electrical signals or piloted hydraulic.

Most subsea valves are fail safe close, i.e. close on loss of power, signals or hydraulics, yet have to be controlled at random to open or close both locally and remotely and even the simple direct hydraulic systems contain some complex hydraulic controls topsides.

The entire hydraulic system will be standard off the shelf equipment, installed within the ICB buoy and will consist of:

1. Hydraulic power units (HPU) consisting of hydraulic storage tank, hydraulic pumps, filters and controls.

2. High pressure storage accumulators with integral valves and gauges.

3. Control panels, one per well.

4. Umbilical riser interface box.

## 6.1   HYDRAULIC POWER UNITS

These will consist of two industrial hydraulic power packs.  Each pack consists of 125 litres of hydraulic fluid storage capacity, at ambient pressure, a 7.5 hp electric motor and starter, pump, return line filter, valves and solenoids, pressure gauges, high pressure filters and heaters.

Typically the size of **each** unit will be:

H = 700mm, W = 570mm, D = 1050mm

Weight approx. = 350 kgs, each (operating)

## 6.2   BLADDER ACCUMULATORS (STORAGE CYLINDERS)

The high pressure hydraulic storage delivered from the power packs will consist of two 345 bar accumulators and three 207 bar accumulators minimum, to satisfy the opening and closing requirements of all valves together on 1 tree.  The accumulators will be delivered complete with shut off safety blocks, gauges and valves, clamps and brackets.  Sizes:  207 bar accumulators - H = 1393mm, dia = 220mm, Weight = 67kg each, 345 bar accumulators - H = 1143, dia = 230mm, Weight = 78kg each.

## 6.3   CONTROL PANELS

There will be up to four hydraulic control panels, one for each tree.  Each unit will have a facia panel containing six handle operated three port rotary valves, eight gauges and identification labels.  The main body of the control panel will contain filters, solenoid valves, hand valves, pressure relief valves, piping and fittings etc. It is anticipated that each panel will have the approximate sizes of H = 750mm, W = 750, D = 1000mm, weight approx. = 150kg.

## 6.4   DYNAMIC RISER INTERFACE BOX

The dynamic Riser Interface Box (DRIB) is used to terminate the flexible riser hydraulic tubes and the riser electrical/signal conductors and to interface these connections with the onboard tubing and wiring installation.  It is proposed that the DRIB is mounted on the deck over the buoys sensor well allowing the riser to come up on deck through the sensor well.  The unit will provide a disconnection point and test point between the riser and buoy.  There will be an additional penetration through the deck to allow hydraulic piping and signal wiring into the bulkhead using a standard transit or similar device.

The enclosure will be to IP 66 within which there will be a junction box for signal connections also rated at IP 66.

## 7.0    ELECTRICAL SUPPLY AND GENERATION

The generators will be up to 10kVA diesel driven air cooled units at 1500 rpm. Units will be dual voltage output with electric start operator by low battery voltage or low hydraulic pressure. The generators are required to provide the following loads, all of which are intermittent. Two Units will be provided on 100% dual redundant basis with automatic changeover.

| Intermittent Loads | |
|---|---|
| Hydraulic pumps | 5.75kW |
| Methanol pump | 4kW |
| Battery Charging | 6kW |
| Heating/Cooling | 500W |
| Lighting | 100W |
| Continuous Loads | |
| Communications | 30W |
| Instruments and Controls | 350W |

Each generator will be capable of supplying the buoy's power requirements under any situation, including full load. Battery charging will be the most frequent requirement of the generator. Sufficient batteries will be installed to run the buoy for 1 week. It is therefore anticipated that the generator will be used at least 12 hours a week.

It should be noted however (that with regard to the hydraulic pump load), that the maximum period of charge up under "cold" conditions will be a maximum of 30 minutes after which time the maximum load will be seen during maintenance or special operations mode. Generally once the accumulators are charged on completion of commissioning the system should remain dormant until the annual maintenance therefore the only electrical load used will be for the communications, instruments and Chemical injection if required plus any heating or cooling if necessary.

Each generator assembly will be a complete package consisting of driver, generator, generator controls, switchgear, instruments, heaters, etc.

**GENERATOR FUEL**

Enough diesel fuel will be stored onboard the buoy to permit as a minimum 12 hour's running time per week for a single generator for one whole year.

The fuel consumption will be monitored and the status of fuel and of the generator sets will be part of the housekeeping functions. Refuelling would normally be carried out during routine maintenance visits to the buoy and facilities will be provided on the ICB to ensure this operation can be carried out quickly and safely.

## 8.0    CHEMICAL INJECTION FACILITIES

With the existing ICB, which is a prototype only, there is sufficient storage capacity for methanol to allow 2 single well startups and 0.7 litres per hour trickle feed injection, to prevent hydrate formation assuming the startup requirement is in the order of 1 cubic metre of methanol for 1 hour. If, however, the storage tanks were recharged after well startup, there would be enough capacity provide up to 2 years supply of methanol and other chemicals.

The chemical injection power will be derived from a dedicated hydraulic accumulator otherwise the metering pump package will be standard offshore equipment.

In cases where the quantity of injection chemicals required is much greater, then seabed storage could be provided, with a supply line connected to the metering package, together with a return line to the tree via the dynamic riser.

## 9.0    DYNAMIC CONTROL RISER

In order to distribute the hydraulic supplies from the standard subsea controls equipment within the ICB to the tree valves and to receive signals from the tree instruments it will be necessary to provide a reliable dynamic multibundle riser.

The riser will require to consist of, as a minimum

> 6 hydraulic ½" hoses
> 3 Chemical injection ½" hoses
> 6 Instrument/signal pairs for downhole and wellhead pressure and temperature transmitters, choke position sensing and downstream pressure transmitter.

There is little doubt that the Dynamic Riser is seen by most Engineers as the weak link in an ICB system, but why? How often do lift cables, site welding cables, crane and coal cutting machinery cables fail.

Many features and papers have been written and published regarding the reliability of umbilicals and the causes and evaluation of failures.   One of the better recent papers "Fatigue, Internal Stresses and Deformations of Electrical Umbilicals" comments that factors affecting fatigue life are the difference of maximum and minimum stresses and the combination of bending and axial forces and most if not all articles "point the finger" at failures being caused by the wrong combination of conductor stranding, insulation and sheathing materials and structural design.  We believe that with the use of the correct and more traditional materials and designs, the dynamic riser will provide many years of trouble free service as our new specification hopes to prove.

Dynamic riser performances have been modelled on water depths as shallow as 26m with currents of 9 m/s up to 150m depth providing a full range of North Sea potential applications.

The reasonable capital cost and the method of installation, makes replacement both practical and economical, should a catastrophic incident befall the riser since the use of divers has been eliminated and the cost of holding a spare riser is not prohibitive.

We envisage that the tree end of the riser would be installed using an ROV, first deploying and then making the final connection with the aid of the riser ROV connectable stab plate as used on DISPS.  After carrying out the usual tests to ensure the integrity of the stab plate connection, the riser installation vessel will lay away lowering the riser, its clump weight and then the mid point buoyancy assembly. Finally the ICB end of the riser would be pulled up into the riser tube and like any umbilical the Dynamic Control Riser will require a hang off and bend restrictor at the ICB end which will be installed as any typical installation.  After hanging off the riser, its pigtails would be connected into the Dynamic Riser Interface Box (DRIB) using industry standard JIC connectors and the instrument pairs terminated into the stainless steel IP66 Exe junction box within the DRIB.

## 10.0   ANCHORING FACILITIES

During their lives as offshore data gathering buoys, the buoy and other similar buoys have been moored in many different locations around the British Isles by THORN; including deep water in the Western Approaches, West of Hebrides, West of Shetland and shallow water off the Norfolk coast.  In each case the buoys were moored for a minimum of three years, the data collected being transmitted back to the Meteorological Office in Bracknell and via a Thorn satellite link.  This proved to be a very reliable system.

*Benjaminsen JT: Brown.N., Due-Andersens. Stafford M. and Walden A.O. June 1992 "Fatigue internal Stresses and Deformations of Electrical umbilicals Experimental Work" Isope 92-B5-01B)

The anchoring system used was a 3 anchor configuration using 1 tonne anchors and 32mm chain. In order to improve the reliability of the anchoring system, for a more hazardous duty and to protect the Dynamic Riser should one anchor fail, it is proposed installing six 1 tonne anchors, this will ensure that, should an anchor fail, the two anchors either side of the failed one will take the load and ensure that the buoys excursion is kept to a minimum.

The following parameters were used in modelling the buoy loads and have been obtained from the Metocean Parameters - supporting document to Offshore Installations : Guidance on Design, Construction and Certification - Environmental considerations, based on a representative 50 year return period at the Alison Field location (54°N, 2°E).

| | |
|---|---|
| Significant wave height | =9.5m |
| Maximum design wave height | =17.7m |
| Maximum design wave period | =14.7 secs |
| Current (depth averaged storm surge) | =0.9 m/s |
| Spring tide amplitude | =1.15m |
| Storm surge | =1.9m |

## SUMMARY OF RESULTS - 50 YR RETURN PERIOD

| Description | Calm Sea | Environmental Load | |
|---|---|---|---|
| Wave direction (°) | - | 0 | 30 |
| Approx. height of wave crest above SWL (m) | - | 8.75 | 8.75 |
| Tension in the mooring line (T) | 2.0 | 26.0 | 18.0 |
| Excursion (m) | 0 | 5.2 | 5.4 |
| Height of clump above seabed (m) | 0 | 1.25 | 0 |
| Ground chain (m) | 140.0 | 25.0 | 60.0 |
| Angle of mooring line to horizontal at buoy (°) | 53.0 | 19° | 21 |

## NOTE:

| | | | |
|---|---|---|---|
| 1. | Horizontal environmental load | = | 27.5T |
| 2. | Clump weight | = | 3T |
| 3. | 1⅝in chain : weight (submerged) | = | 335Kg/m |
| | min breaking load | = | 140T |
| 4. | Pretension in mooring line | = | 2T |
| 5. | Total payout length | = | 190m |
| 6. | Water Depth: | = | 28m |

| | | |
|---|---|---|
| SWL | = | 26m |
| Wave Crest + tide and surge | = | 27.9m |
| Wave trough - tide | = | 16.0m |

## INSTALLATION VESSELS AND PROCEDURES

It is believed that the ICB and its anchoring system can be installed relatively easily by a RSV given the following weights:

anchor and chain + lump weight 10.5 Tonne

Dynamic Riser plus accessories 17 Tonne
(inclusive of 2 Tonne lump weight)

Previously the Buoys have been towed out to site but should they require lifting by the RSV the weight is:

| | |
|---|---|
| Buoy | 22 Tonne |
| Equipment | 4.2 Tonne |
| Fuel and Chemicals | 7.5 Tonne |
| Total Wet Weight | 33.7 |

## 11.0  OPERATION AND MAINTENANCE

One of the major benefits of the ICB system is the ease of maintenance and relatively low expense. With all the controls equipment mounted within the Buoy the use of divers for repair or replacement is not necessary and access and methods used would be no different than working on board a boat.

All equipment specified is standard subsea and offshore design thereby alleviating the need for special skills or training. The Buoy would be serviced via a supply boat or small access vessel depending on the weather.

All generating, hydraulics, communications and controls equipment etc. are duplicated and with its onboard management and DCS system, repairs or replacements can be scheduled for planned shutdowns or during periods of good weather.

From the Open estimate of planned activities two visits a year have been anticipated for the top up of diesel fuel oil, chemicals for injection, filter changing and general inspection.

For its 20 year life we have estimated pumps being replaced every 3 years and generators and electronics every 4 years. The total planned OPEX is estimated at 62% of the total OPEX of £1.732m over the 20 year period. It is anticipated that a major overhaul would take place every 4 to 5 years at which time the pumps and generators would be stripped down and replaced as necessary. The dynamic riser and moorings would also be replaced.

Preventative maintenance and analysis can be carried out at any time (subject to weather conditions) thereby providing the user with a greater comfort feeling and more flexible field management.

## 12.0  RELIABILITY

Most of the existing electronic equipment will be removed from the buoy and minor modifications carried out to the internal structure making it suitable for its new use, the external structure will not be altered, apart from these changes the buoy will be the same as that successfully used in the North Sea for 5 years. One of the reasons for the reluctance in using such a system for well control in the past is the possibility of damage to the Buoy and hence shutdown, or worse, the non-availability of system shutdown.

It is on record that during their period of operation offshore, neither of the two Buoys failed to communicate even though each buoy was struck by vessels at high speed.

It is felt that with additional visual and audible beacons the chances of collision would be reduced and with impact sensors on board the buoy, should such an event occur, then the Master Control Station would be informed and an inspection arranged to check for damage.

It should also be noted that should the outer plates of the buoy be penetrated in a collision the water tight bulkheads will prevent water entering the communications equipment area.  As well as the communications system, the power generation and battery supplies will be dual redundant, the batteries being compartmentalised on opposite sides of the buoy within bulkheads giving maximum protection and reliability.

### Securing on Seabed

As discussed in the section on Anchoring (10.0) 100% has been added to the capacity of the anchoring system over that used for the successful data gathering programme.

### Communications and Hydraulics

As previously mentioned, all controls equipment will be standard subsea/offshore high reliability specification, providing familiarity and availability to the user.  The hydraulic and electrical supplies together with the communications equipment will be dual redundant with automatic transfer in event of failure.

All controls parameters will be continually monitored and any changes, ie. out of limit readings will be actioned by the buoy.  Failure to respond in this way, within a specified time, will cause the ICB's computer to take any action necessary ie, shutdown.

### Dynamic Controls Riser

At the time of writing the riser Purchase Specification has not been completed but with the assistance and recommendations of leading umbilical and subsea cable manufacturers, it is felt an ideal system design can be achieved to last at least 10 years.  Again, hydraulic leakage will be constantly monitored together with earth leakage and line earth loop impedance reading on the signal cables. During planned shutdown full advantage would be taken to carry out any fault finding tests or inspection on the Riser.

### General System

A full Reliability Analysis will be carried out during detailed design and the results published.

## SUMMARY

This brief paper outlines the concept for a new method of controlling subsea wellheads.

The system uses only proven technologies. It is the combination of these technologies into a novel configuration that leads to the benefits described previously. These may be summarised as:

Low cost installation

High reliability, remote operation

Intrinsically safe, autonomous operation

Re-usable

Diver-free, low cost maintenance

A cost study has been carried out comparing the ICB with a conventional system for a single well with a stepout distance of 15KM demonstrating cost savings in the order of £2.5m for a manifold system and longer stepout distances even greater savings can be made.

The conceptual design of the system has been completed and the project is now ready for implementation. Installation of the first system should take place during summer 1995.

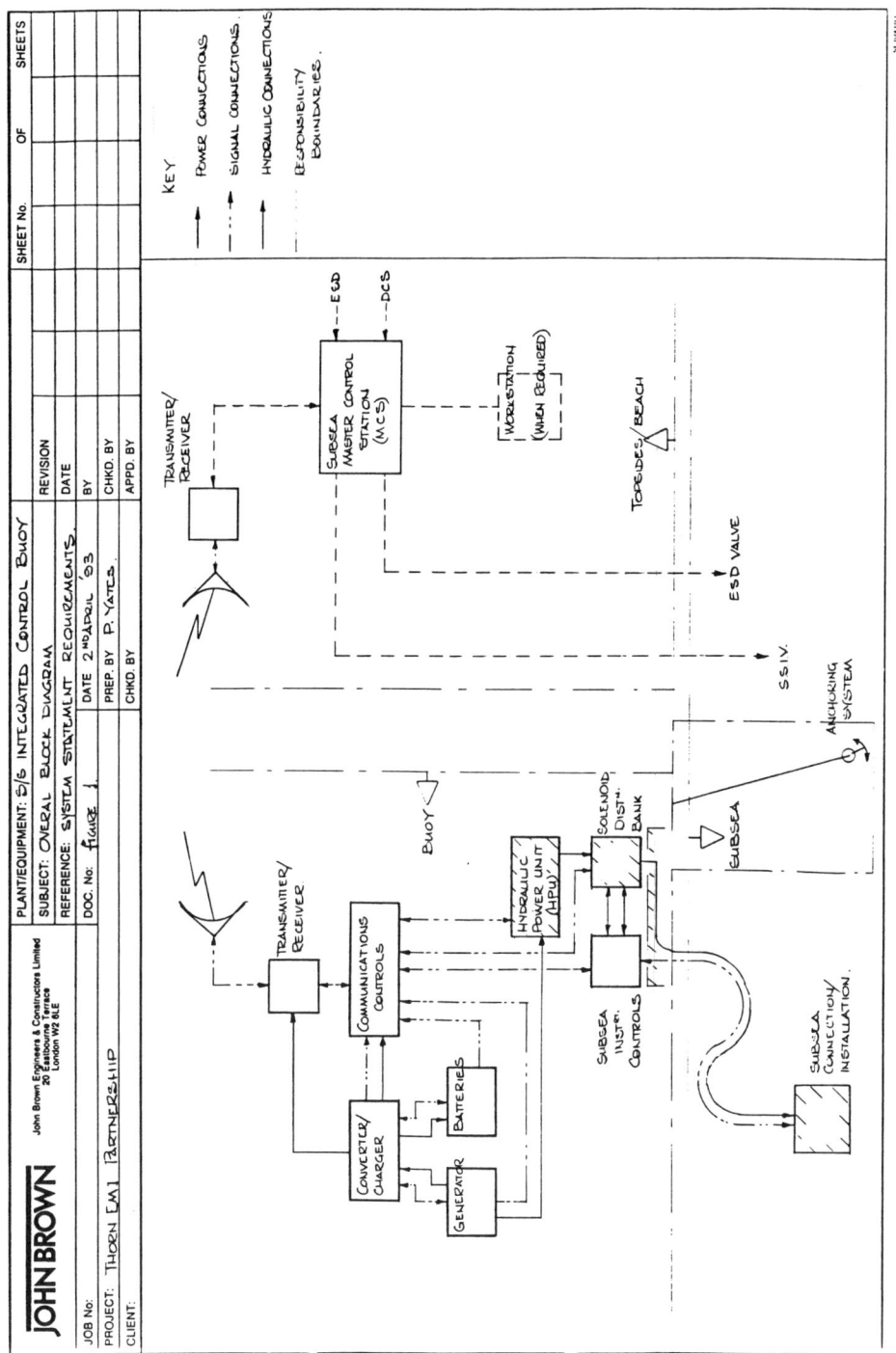

# SPARCS autonomous control system

M C THEOBALD
Research and Development Group
FSSL Ltd
Monarch House
Victoria Road
London W3 6UL
United Kingdom

## ABSTRACT

This paper describes a Subsea Powered Autonomous Remote Control System (SPARCS) designed to control subsea wells without the use of electro-hydraulic umbilicals between the seabed and topsides. SPARCS is intended to provide a low cost control solution to justify subsea developments in small and marginal oilfields.

Each of the main system components is described and a commercial justification is given for the system as a whole.

SPARCS is highly innovative in that it has been designed to control hydrocarbon and water injection wellheads without the use of a control umbilical. It relies on acoustic telemetry, with seawater as the transfer medium. The system's electrical power subsea is produced by a turbine generator fitted to a water injection flowline or a thermo-electric generator fitted to a production flowline. Hydraulic power for operating wellhead and downhole safety valves is produced from a subsea unit.

Development is well advanced and will culminate in an 18-month offshore application trial, planned to commence in 1995 on the Saltire development in the North Sea.

## INTRODUCTION

Offshore hydrocarbon recovery using subsea production facilities is now a mature technology and is used extensively as North Sea oil exploration moves towards marginal fields and smaller fields connected to the existing infrastructure. In order to justify these developments there is continuous pressure to reduce costs.

The SPARCS system has been designed to meet these cost reduction targets. SPARCS (Subsea Powered Autonomous Remote Control System) is a system designed to control subsea development wells without the use of electro-hydraulic control umbilicals back to the surface. The umbilical is one of the main cost items of a subsea control system, and SPARCS is targeted at providing maximum cost-reduction benefits for single well or satellite wells conventionally requiring dedicated umbilicals of lengths up to 10 km.

*Volume 32: Subsea Control and Data Acquisition, 155–170.*
© 1994 *Society for Underwater Technology. Printed in the Netherlands.*

SPARCS has two groups of components:

- A Subsea Control Unit mounted near the well and providing all the control and data-monitoring functions. Depending on the type of well to be controlled, electrical power is produced by a turbine generator driven by a water injection flowline or a thermo-electric generator clamped to a production flowline. The unit also contains a means of generating hydraulic power on the seabed using a closed-loop-reservoir hydraulic power unit and a battery system.

- A Surface Controller comprising an operator-interface console and the acoustic telemetry system.

A diagram of SPARCS is shown in Figure 1 and can be compared with the typical existing multiplexed electro-hydraulic control system using umbilicals shown in Figure 2.

## DESIGN PHILOSOPHY

The overall design philosophy of SPARCS has been to produce an autonomous subsea production control system that conforms to the same functional requirements, engineering standards and safety codes of practice that existing control systems satisfy. Because it is an autonomous system, particular emphasis has been paid throughout the development on safe operation; any single point failure will not leave the well in an unsafe condition.

Due to the difficulty and cost of access to subsea systems, the design has been based on the need to ensure high reliability and availability. The system needs to be inherently reliable, using only components of high quality and ones which, wherever possible, have a proven or field-tested history.

The target for availability of the SPARCS system is 98% with a down-time of 10 days in a year. These figures may not be easily verified, since failure to meet this target could be due to environmental conditions beyond operational control (eg acoustic disturbances).

The materials used in the system have been selected on the basis of their resistance to deterioration during long-term immersion in seawater and in contact with the operating well and hydraulic fluids. The design will minimise the possibility of electrolytic action due to dissimilar metals. Cathodic protection systems are to be developed in conjunction with the protection system used on adjacent wellhead subsea equipment.

The design life of the subsea units of a SPARCS system is set at 20 years, with periodic maintenance. For application on a water injection well, the system will be designed to operate for a minimum period of at least 2 years without diver intervention for maintenance. As the electro-hydraulic umbilical tends to be one of the most unreliable components of conventional control systems, the SPARCS system will increase overall reliability even with the additional subsea-deployed hardware.

Incorporation of new technologies into SPARCS, such as electrical actuators, seawater composite actuators and seawater batteries[1,2], will all be evaluated and considered as

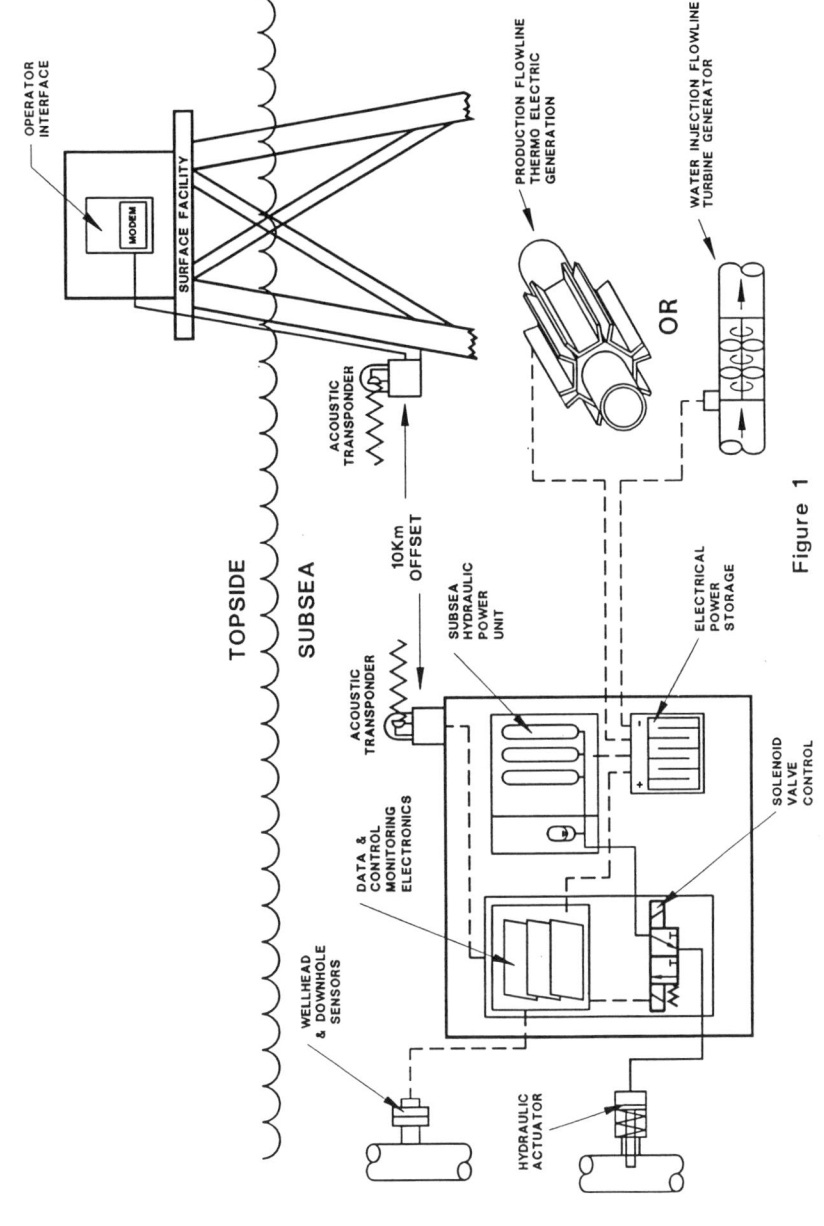

SPARCS - BLOCK DIAGRAM

(SUBSEA POWERED AUTONOMOUS REMOTE CONTROL SYSTEM)

Figure 1

EXISTING MULTIPLEXED SUBSEA CONTROL SYSTEM - BLOCK DIAGRAM

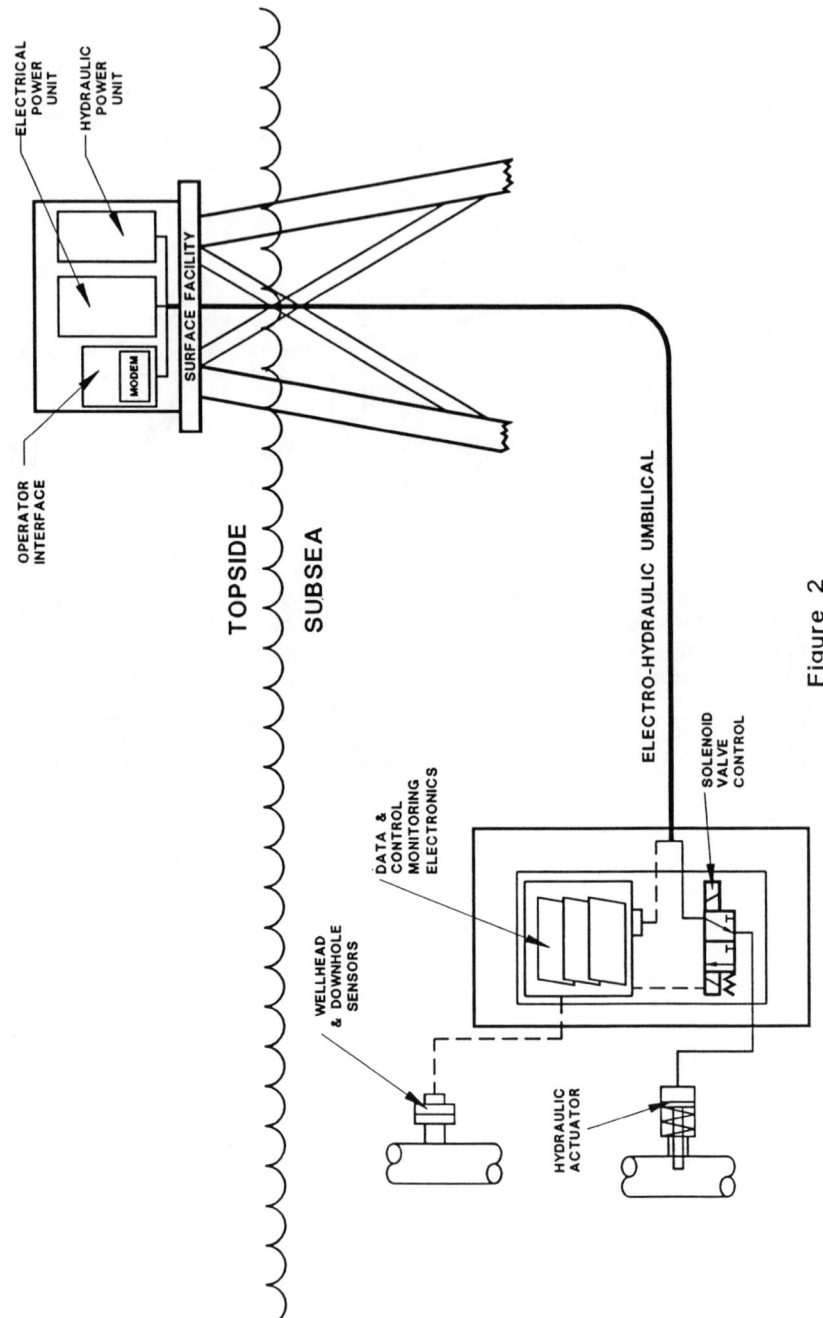

Figure 2

qualified equipment becomes available. Preliminary consideration has already been given to their use in the system.

As SPARCS is autonomous and generates the power it needs on the seabed, it does not require any equipment at the surface on the platform facility for the electrical and hydraulic power units used in traditional control systems. Removing these power units from the hazardous environment found on offshore platforms will result in increased safety levels for personnel and operations.

## COST JUSTIFICATION AND SAVINGS

The cost savings from using SPARCS are not just in the elimination of the umbilical but also from the reduction of topside equipment, which is often situated in areas with hazardous-operation environments.

The typical cost savings of SPARCS compared to a conventional control system with a single well at an offset distance of 7.5 km is estimated at nearly £1500k depending on exact field conditions.

Figure 3 shows a cost comparison between SPARCS and a conventional umbilical system for a single-well development at offset distances up to 10 km. The conventional topside equipment is costed at £300k coupled with a subsea control module at £250k. Umbilical manufacturing cost is estimated at £100k per km (£100 per m) with £100k per km installation costs.

The cost of a SPARCS system is estimated at £1200k, depending on the exact application, and the cost is independent of offset distance. (Note that items common to both conventional and SPARCS technology have not been included in this cost comparison.)

Figure 3 shows that the break-even offset distance is around 2.3 km, with steadily increasing cost advantages using SPARCS up to 10 km. The maximum offset distance is currently limited by the physical constraints of the presently available acoustic communication system, although developments taking place in non-umbilical communications technology are expected to increase this distance to around 25–30 km.

## PROJECT STATUS

As part of a joint industry programme, Enterprise Oil plc, Elf Enterprise Caledonia Ltd and FSSL Ltd are combining in a demonstration project to fully qualify SPARCS in realistic offshore conditions. In addition to the SPARCS participants' financial contributions, support has been successfully obtained from the European Commission Thermie programme. This aims to stimulate innovation by encouraging the demonstration phase of potential energy-technology projects.

The SPARCS development project started in 1989, with the concept design completed over a period of two years as shown in Figure 4. Earlier work on subsea autonomous control systems is reported in Reference 3. The initial hardware manufacture and test phase started in 1992 and concentrated on the items with the highest R&D content.

SPARCS
CONTROL SYSTEM COST COMPARISION
1 WELL SUBSEA DEVELOPMENT
TYPICAL COSTS
EXISTING E/H SYSTEM
SPARCS SYSTEM
BREAKEVEN DISTANCE APPROXIMATELY 2.5km
0
1
2
3
4
5
6
7
8
9
10
PLATFORM TO WELLHEAD DISTANCE (Kilometres)

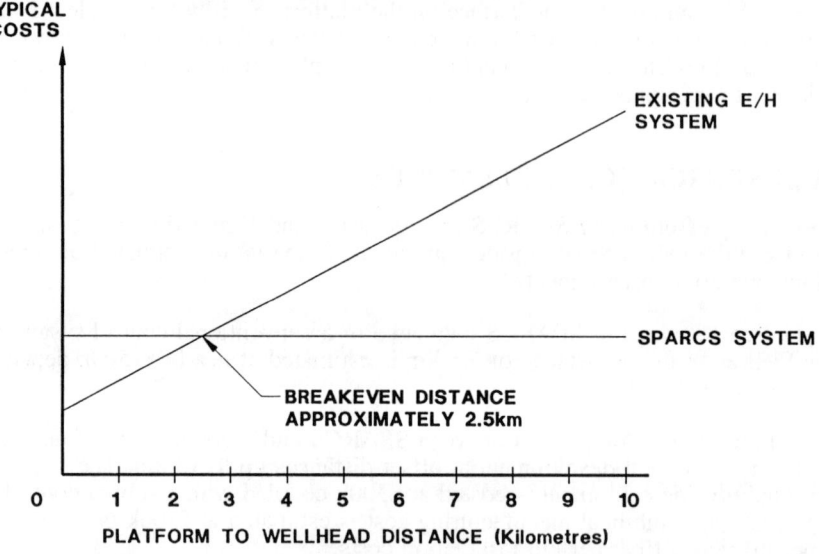

Figure 3

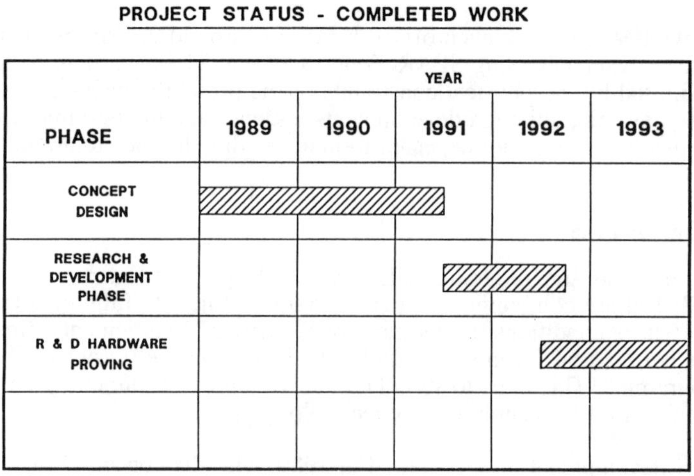

PROJECT STATUS - COMPLETED WORK
PHASE
YEAR
1989
1990
1991
1992
1993
CONCEPT DESIGN
RESEARCH & DEVELOPMENT PHASE
R & D HARDWARE PROVING

Figure 4

These were the flowline thermo-electric generator, the subsea hydraulic power unit and the electrical power conditioning/motor control electronics. This work has now been completed with all relevant items successfully tested on land.

The balance of hardware design and testing is due for completion by the end of 1994, and will be followed by an inshore underwater trial involving 3–4 months of continuous operations followed by an 18-month offshore application trial, all as shown in Figure 5.

One of the project participants, Elf Enterprise Caledonia Ltd, has provisionally made available a subsea water injection well in their Saltire development for the offshore application trial. This North Sea trial will extend over more than a one-year period to evaluate the system over the full range of seasonal weather variations. This is particularly important in order to appreciate fully the range of possible effects on the acoustic communications.

Although the equipment to be used in the application trial will generate electrical power from the turbo-electric generator fitted to the water injection flowline, the thermo-electric generator will be deployed so that it also can be monitored to assess its long-term performance.

## DETAILED SYSTEM DESCRIPTION

SPARCS will have the ability to control a single remote water injection or production well. It will do this through two groups of components: the Surface Controller and the Subsea Control Unit. The Surface Controller is the control station on the topside of a platform and provides the operator with the means of controlling and interrogating the subsea system. The Subsea Control Unit is the outstation that facilitates all the control and data monitoring functions at the well.

### Surface Controller

The following equipment will be installed on a platform to comprise a typical SPARCS Surface Controller:

- an operator control console installed in the equipment room on deck. It has an operator keyboard, a display monitor, a printer and a functional control panel station. The computers are rack mounted and include CPU, memory, I/O cards, disc storage and the computer watchdog.

- an acoustic transponder to house the communication modem, the acoustic transmitter-receiver and an interface to a directional hydrophone. It is currently proposed that the transponder is clamped and bolted into position, although consideration is being given to a wireline system to allow its recovery to platform deck level. Alternatively, it could be mounted on a separate cage frame on the seabed and directed towards the Subsea Control Unit.

- a power and communication cable to connect the operator control console with the transponder on the jacket leg (via an acoustic modem)

- a power supply for the operator control console and the acoustic communications system
- an interface cable to the Platform Emergency Shutdown system.

## Subsea Control Unit

The following equipment will be installed at the Saltire subsea structure for the 18-month offshore application trial. Together it comprises the Subsea Control Unit:

- subsea control module
- subsea solenoid control valves
- hydraulic power unit
- turbo-electric generator (TEG)
- (thermo-electric generator (THEG))
- submersible battery system
- power conditioner system
- motor controller system
- subsea electronics module
- subsea acoustic transponder
- pressure sensors and associated subsea cables and connectors
- intervention unit to enable well control to be switched to operate via the existing direct hydraulic platform link or the SPARCS equipment.

The *subsea control module* (SCM) supports the other sub-systems in the list above in making up the entire Subsea Control Unit. It is the SCM that is mounted on the support frame near the wellhead.

The *solenoid control valves* are the electrically operated control valves used for actuating the tree and downhole valves, and are contained within the SCM.

The *hydraulic power unit* is required to provide hydraulic fluid power for the valve control system. It includes reservoir, pumps, motors, accumulators and filters. Dual high-pressure supply lines will be fitted.

The *turbo-electric generator* (TEG) converts the kinetic energy of a water injection flowline into electrical energy to power the Subsea Control Unit. It will be fitted when SPARCS is controlling a water injection well.

The *thermo-electric generator* (THEG) will be mounted on a nearby production well flowline. It will be interfaced to SPARCS during the offshore application trial solely for long-term experimental evaluation. In a fully operating system, it would provide power when SPARCS is used to control a production well.

The *battery system* stores the electrical energy from the subsea power source in periods of low power demand and releases it to meet peak demands for electrical power (ie motor operation and acoustic communications).

The *power conditioner system* interfaces the THEG/TEG, the electronics and the batteries. It provides power to the electronics and it maintains the batteries' charge.

The *motor controller system* provides the interface between the power source and the hydraulic pump motors. It provides a means of motor control.

The *subsea electronics module* contains the electronics for controlling and monitoring valves and data. This unit is microprocessor controlled and contains local control systems to operate and, if necessary, shut down the well without surface intervention.

The *subsea acoustic transponder* houses and interfaces a communications modem, an acoustic transmitter/receiver and a directional hydrophone. It will be connected to (or near) the wellhead Subsea Control Unit and will be directed towards the acoustic transponder on the surface to ensure a clean line of sound.

## Typical well and wellhead data

In the offshore application trial, the subsea control system is likely to operate a wellhead completion with the following characteristics:

- single water injection well       between 2 and 10 km from the surface facility
- completion type       5 in and 2 in dual bore
- design pressure (hydraulic)       517 bar.

The water injection well has the following complement of valves:

- tubing master       $5^1/_8$ in
- tubing wing       $5^1/_8$ in
- tubing choke (where applicable)   4 in
- subsea safety valve (DHSV)       $5^1/_2$ in.

The valves have the following typical characteristics:

*$5^1/_8$ in valves*
- swept volume       3.2 litres
- typical open pressure       54 bar
- typical cracking pressure       47 bar
- maximum working pressure       207 bar

*DHSV tubing valve*
- swept volume       0.09 litres
- typical pressure       517 bar (at 345 bar well pressure)
- typical cracking pressure       (not available yet)

*choke valves (for information)*
- body pressure       345 bar
- actuator range       0–100%       0–90 steps
- displacement       1.59 litres       (90 steps)
- actuator pressure       57 bar.

**PROJECT STATUS - FUTURE WORK**
**DESIGN COMPLETION, INTEGRATION & TEST**
**INSHORE TRAILS**
**OFFSHORE ASSEMBLY/COMMISSIONING & MONITORING**

| PHASE | YEAR | | | |
|---|---|---|---|---|
| | 1993 | 1994 | 1995 | 1996 |
| DESIGN COMPLETION | ▨ | | | |
| MANUFACTURING /TEST | | ▨ | | |
| INSHORE TRIALS | | | ▨ | |
| OFFSHORE COMMISSIONING | | | ▨ | |
| OFFSHORE MONITORING | | | ▨▨ | |

Figure 5

The tree valves to be controlled are master, wing, SSSV, choke valve open and choke valve close with operating pressures up to 517 bar. A sixth (spare) control function will also be implemented.

The water injection well has the following data sensors:
- tubing pressure           0–400 bar   4–20 mA
- annulus pressure          0–400 bar   4–20 mA
- choke differential pressure   0–20 bar   4–20 mA
- choke position            0–100%   4–20 mA.

Flow from a production well can be calculated from the differential pressure across the choke and the valve's Cv position curve. For a water injection well, flow can be determined from the TEG output voltage.

A facility will be added, if practical, to monitor downhole pressure and temperature from a nearby producing well. Typical downhole pressures and temperatures are up to 690 bar and 100°C.

## Power generation

The turbo-electric generator (TEG) will be installed in the flowline spool of the water injection well. It will convert the kinetic energy of the injection water into electrical energy providing continuous power to the entire Subsea Control Unit. It will be designed to operate to 250 m depth in a seawater environment. In general, it will provide at least 100 W of output power. Output will be three-phase low voltage AC (11 V to 20 V). The output voltage of the TEG may also be exploited as means of measuring water flow in the pipeline (see above).

A thermo-electric generator (THEG) generates power by relying on the temperature differential between the hot produced fluids of a production well and the cold surrounding seawater. It also will be designed to operate to a minimum 250 m depth in a seawater environment. The prototype THEG will provide up to 20 W of power and output will be a low voltage DC (3 V to 8 V). Nominal power of the THEG when fitted to a production flowline with a differential temperature of 50°C will be 100 W.

Figure 6 shows the prototype THEG at pre-encapsulation stage.

## Acoustic transponder (surface and subsea)

Each acoustic transponder (one on the platform and one at the wellhead) will consist of:
- a directional transducer (hydrophone)
- a transmit unit
- a receive unit
- a transmit/receive switch

- an acoustic modem with input/output to the surface operator's console or the subsea electronics module (SEM) respectively
- local power supply conditioning.

The surface and subsea acoustic transponders between them are required to carry out all communication functions of SPARCS. They will provide secure acoustic communication using a semi-duplex serial link with error detection and correction facilities. In normal use, the subsea unit will act as a slave to the surface equipment, although under certain circumstances the subsea unit may initiate a transmission to indicate fault conditions detected within the Subsea Control Unit.

The specification criteria for the transponders are:

- system range                     10 km
- data resolution                   12 bit
- host I/O                          RS422
- low power consumption             200 W maximum
- target error rate                 better than 1 in $10^4$
- data rate                         better than 40 baud.

## Hydraulic power unit

The hydraulic power unit is to be a closed loop system where all leakage and venting flow is returned to the reservoir. The reservoir is to be fully pressure compensated to ambient sea depth pressure. It will be sized to supply all the tree actuators and the accumulators at once with an allowance for irrecoverable hydraulic fluid leakage over the maintenance period. The unit will provide hydraulic power at two pressure levels.

Return line surge accumulators are to be used to limit the peak flow into the reservoir/ receiver system. The reservoir receivers will have provision for the separation of gas and solids from the return flow and a means of venting gas pressure build-up. It is essential that an impermeable interface is introduced between the hydraulic oil and the pressure-compensating fluid. The suction line will incorporate a suction strainer to prevent large particulate contamination reaching the pump.

The pumps are submersible DC-motor driven close-coupled pump sets powered through the motor controller system. Both motors and pumps are immersed in an oil-filled and pressure-compensated environment.

Specification requirements of the components of the hydraulic power unit include:

- receiver capacity                 100 litres
- reservoir capacity                300 litres
- reservoir vent pressure           0.5 bar above sea pressure
- pressure nominal                  517 bar (high pressure)
- pressure nominal                  207 bar (low pressure)
- suction line strainer             100 μm
- flowline filtration               3 μm

- motor pump voltage           nominal 60 V DC
- motor pump output power
  (70% efficiency) LP        350 watts minimum
  (70% efficiency) HP        300 watts minimum
- flow LP                  0.7 litres/min minimum
- flow HP                  0.2 litres/min minimum.

SPARCS uses a closed-loop hydraulic system which does not eject control fluid to the sea during well valve operations. Although the majority of control fluids are water-based, the use of a closed-loop system inherently has less impact on the environment. A considerable amount of this fluid is used in offshore hydrocarbon production and it is estimated that SPARCS will reduce global pollutant emission by approximately 3000 litres per annum.

### Hydraulic fluid

The hydraulic fluid to be used in the Subsea Control Unit will be a water-based fluid compatible with a typical system fluid (HW540) but with a higher viscosity – similar to that of a mineral-oil-based fluid. The higher viscosity should reduce the amount of system fluid leakage.

Oceanic HV355 will be used as the control fluid in the offshore application trial. The suppliers (Marston Bentley) have confirmed it is chemically similar to HW540 and is fully compatible. Its characteristics are:

- kinematic viscosity        60 cSt at 0°C    to    29 cSt at 20°C
- density                    1065 kg/m$^3$.

### Submersible battery system

The battery system is required to provide back-up power for periods of peak power demands. It will be installed and operate (including charging) in an oil-filled environment at depths up to 250 m. A pressure-equalised battery system using flooded liquid-electrolyte lead acid cells will be used. The nominal battery system capacity will be 18,000 watt hrs at 20°C. Nominal system voltage is to be 60 V DC with a maintained float charge. The submersible battery system will be integrated with the other components of the SPARCS system.

### Power conditioner system

The power conditioner system provides the interface between the subsea generators (TEG or THEG) and the electronics/battery system. It will power the Subsea Control Unit as a priority over all other system power requirements. It will also charge the battery system at float and higher charge rates.

It is intended that the power conditioner circuit-board along with the TEG or THEG interface and low voltage detect and disconnect (LVDD) unit be installed within a one-atmosphere, nitrogen-filled pressure vessel – called the power conditioner module. Components from the motor controller system (see below) are also fitted in the power conditioner module.

## Motor controller system

The motor controller system provides the interface between the batteries, the power source of the motors, and the motors. The batteries will provide the source of energy for the motors via motor power units. Motor control electronics will interface with the system microprocessor and motor control pressure switches to the motors, enabling on/off control.

The motor power units and motor control electronics will be installed within the power conditioner module, alongside components of the power conditioner system.

## Subsea electronics module

The subsea electronics module (SEM) is responsible for:
- acoustic communication control (subsea module)
- mode operation
- operating the hydraulic control valves (to open or close well valve actuators)
- exciting and monitoring the system sensors
- monitoring various system voltages
- data logging of sensor and voltage information
- executive local control actions
- calibration checks
- dealing with loss of communication
- motor control
- down-loadable applications software
- low power consumption (power-down control)
- quiescent receive mode.

The SEM will be responsible for controlling all subsea acoustic transponder transmissions. It will operate in a quiescent receive mode awaiting transmissions from the surface acoustic transponder and will also have the capability to wake up all components of the subsea electronics module. The communication protocol between the SEM and subsea acoustic transponder is to be over an RS422 serial link.

## Mode operation

The water injection well will be operated in six (out of the total of eight) well operating modes. The eight modes are:
- field well shut-in Class 1
- process well shut-in Class 2
- well start up
- normal injection
- (reserved for production well)
- (reserved for production well)
- platform manual override
- well dormant.

Each operating mode has particular requirements for valve functioning. The principle behind mode operations overcomes the limitation that acoustic communications are slow (because of propagation and baud rate limitations). A single mode command from the surface defines a complete tree configuration, replacing several separate valve commands. Mode operation also offers the important feature of savings in subsea power, since the subsea control module needs to acknowledge only one mode message.

If none of the predefined modes are suitable on any one occasion, single valve operation is also available (Mode 7).

### Maintenance

The period between planned maintenance of SPARCS is to be the same as that for other aspects of well maintenance. In general, this is two years for an injection well. For the offshore application trial on the water injection well, the system will be designed to operate for a minimum period of at least two years.

Maintenance of the Subsea Control Unit will involve diver intervention. It will be effected by firstly retrieving the support frame which locates the Subsea Control System to the surface. Maintenance operations will then be carried out. The permanent guide base (PGB) will not be retrieved. Hydraulic fluid replenishment will be treated differently and will not require the need for retrieval of the subsea equipment.

The components of the system that require maintenance are:

*Hydraulic*
- replenish hydraulic fluid          2 years
- replace filters                    5 years
- maintain/replace motor pump sets   5 years
- re-charge accumulators             5 years

*Electrical*
- replace batteries                  2 years

*Acoustic*
- maintain transponder               5 years

*Turbo-electric generator*
- turbine, bearings and seals        5 years.

## ACKNOWLEDGEMENTS
FSSL wishes to thank Enterprise Oil plc and Elf/Enterprise Caledonia Ltd for their support and help in progressing the SPARCS project. Figure 7.

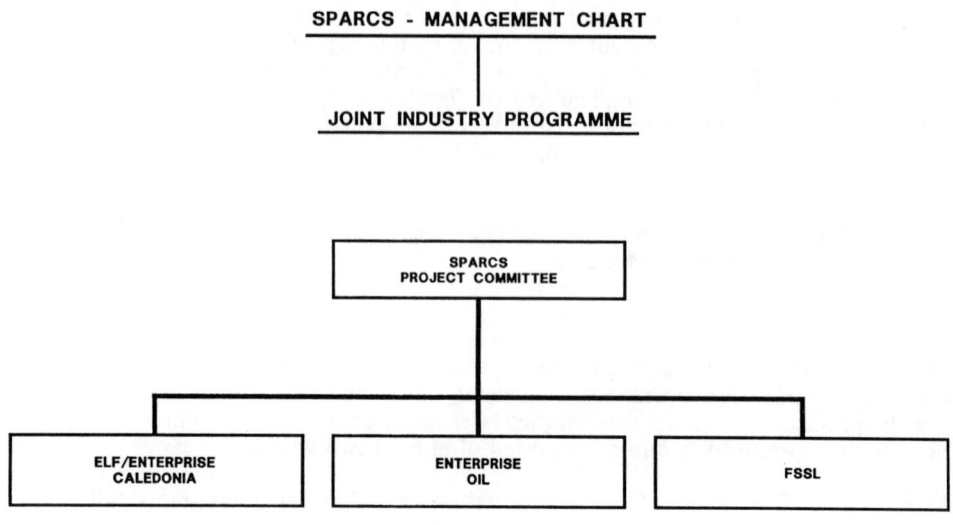

Figure 7

## REFERENCES

1. Loth, W D and Walker, C A (1993) 'Low-cost autonomous subsea water injection systems', in Advances in Underwater Technology, Ocean Science and Offshore Engineering Series, Vol 30: Subsea International '93, Published for the Society for Underwater Technology by Kluwer, Dordrecht, NL, pp 1–12.

2. Tseung, A C C, Shen, P K and Kuo, C (1992) 'Development of an aluminium/seawater battery for subsea applications', Core Research Programme Report No 92/11, Petroleum Science and Technology Institute, Aberdeen UK.

3. Chimisso, C, Dal Lago, C and Franceschini, G (1989) 'Subsea well control system without umbilicals: Performance of the industrial prototype on a field test', in Advances in Underwater Technology, Ocean Science and Offshore Engineering Series, Vol 20: Second Generation Subsea Production Systems, Published for the Society for Underwater Technology by Kluwer, Dordrecht, NL, pp 203–215.

## BIBLIOGRAPHY

Theobald, M C (1992) 'Autonomous control system (SPARCS) for lowcost subsea production systems', in Underwater Technology, Jnl of the Society for Underwater Technology, Vol 19 No 2 Summer 93.

# MULTI-LAYER HOSE LINER OF LOW PERMEABILITY MATERIALS FOR USE IN CHEMICAL INJECTION AND CONTROL UMBILICALS

P W Deane & F Marcoz
DuPont (UK) Ltd
Maylands Avenue
Hemel Hempstead   HP2 7DP
England

## ABSTRACT

The need for improved materials for control and chemical injection liners in umbilicals has been recognised for some time.  The materials, which have been in use for many years, NYLON 11 and HYTREL* both have some limitations in the two critical areas of a) long term chemical compatibility with hydraulic fluids and injected chemicals, and b) permeation of methanol.

As a result of the industry's desire to overcome these short comings, work has been continuing for some time on alternative compositions which would address the problems.

This paper describes some new materials and hose liner designs which have been developed, and which are currently being evaluated in finished hose constructions.  In particular, two multi-layer designs have been developed which appear to offer significant improvements over existing hose liner materials.

## INTRODUCTION

High pressure thermoplastic hoses have been in use for more than 15 years in umbilicals for hydraulic control and chemical injection services to subsea wellheads.  The length and number of these umbilicals has been increasing steadily as technology has developed towards the adoption of more subsea completions at greater distances from the mother platform.

As the use of umbilicals has increased, so has the attention paid to the design and testing of the umbilicals and their component hoses. The experiences of past installations, as well as the analysis of results from many laboratory tests which have accumulated over the years, has led to a much greater understanding of the critical requirements of umbilical hoses, and in particular the hose liner (core tube) constructions.  These requirements can be categorised in the following way:-

o    Long term compatibility with a) hydraulic (control) fluids
                                  b) injection chemicals
                                     (including methanol)

o    Lowest possible permeation rate with respect to fluids, in
     particular methanol

o    Excellent long term mechanical properties, including resistance
     to creep.

o    Cleanliness - in terms of non-contamination of the fluids due to
     extraction of components from the liner material.

In the past, two materials were exclusively used for hose liners:
Nylon 11 and polyester elastomer (HYTREL*).  Each of these materials
has its strengths and weaknesses in the areas of chemical
compatibility; in particular, Nylon 11 tended to be favoured where
water/glycol fluids were commonly used,  whilst HYTREL* had its
strengths in the hydrocarbon based fluids.  However, it has become
clear that the industry demands greater guarantees of long term
overall compatibility to a wide range of chemicals, as well as much
reduced barrier properties with respect to methanol permeation.


## CHEMICAL COMPATIBILITY

It is not the purpose of this paper to go into detail about
compatibility of hose liners with the wide range of control and
chemical injection fluids that are used in umbilical hoses.  Suffice
it to say that the major control fluids used today tend to be
water/glycol based, with some mineral oil based fluids still evident;
also the range of injection chemical used can extend from relatively
inert fluids to some of the more aggressive anti-corrosion and
de-scaling chemicals.  In addition, methanol is injected in large
quantities via umbilical hoses into the wellhead, for flushing
operations.

The nature of the chemical interaction of such fluids with the
thermoplastic hose liner material may occur in several ways:

o    Swelling of the liner material caused by absorption of the fluid
     into the polymer matrix.  This effect occurs quite rapidly
     (usually within a few days, or even hours from first exposure).
     The resulting softening of the liner is not necessarily
     detrimental, unless it is severe, and is usually reversible
     should the fluid be removed.

o    Shrinkage (negative swell) of the liner caused by extraction of
     products (e.g. plasticisers) from the polymer material - not
     usually reversible.  This can result in liners becoming hard or
     perhaps brittle in extreme cases.

o    A combination of both the above, resulting in replacement of
     some of the extractable materials by absorbed fluids, with
     little or no observed change in the material flexibility or
     hardness.

o    Chemical "attack" on the polymer chains or "backbone" by the
     fluid or chemical constituents.  This is potentially more
     serious, as it will in extreme cases result in a loss of
     elasticity and possible cracking of the hose liner over a period
     of time.  The nature of this "attack" may not be evident in the
     early stages, but only after the elasticity has dropped to below
     a certain critical level (depending on material stresses due to
     hose pressure, dynamic flexing, etc).  This type of attack can
     often be accelerated when the material is under mechanical
     stress, particularly cyclic stress.

     "Chemical compatibility" can generally be predicted by the
     results observed from suitably designed laboratory tests.  These
     test procedures have been developed by the raw material
     suppliers, the hose manufacturers, and others over recent years,
     and have made an important contribution to the understanding of
     compatibility of hose liner materials.


PERMEABILITY TO METHANOL

Due to the undesirability of methanol permeability through the hose
liner, and thereby reaching the platform end of the umbilical
(through a "wicking" effect along the fibre braid, assisted by
hydrostatic pressure), or causing cross-contamination of adjacent
hose fluids, this issue has recently been highlighted as a major
concern in umbilical design.

Tests on both established hose liner materials, NYLON 11 and
polyester elastomer, showed that such permeation over the length of a
typical umbilical of several kilometers could amount to several
litres per day.  Although more recently designed umbilicals have
overcome the problem to some extent (by, for example, allowing
permeated methanol to disperse freely in the seawater environment,
or by arranging for collection of the methanol at the platform end),
there remains a desire on the part of operators to eliminate, as far
as possible, the original permeation effect.

As a result, several laboratories and hose manufacturers have set up tests to compare permeabilities of hose liner materials. Some measurements have been made on complete hoses, under pressure, whilst others have been made on the core tubes alone. Although differences exist between numerical results, depending on whether samples were in the form of complete hoses or just core tubes, or were pressurised or not, the results are more or less comparable. They show that the most significant factors affecting permeation of methanol are a) type of liner material and b) temperature.

Some typical results for both nylon 11 and HYTREL*, from a half inch bore hose, at 20°C, under atmospheric pressure are given in the following table:

| LINER MATERIAL | PERMEABILITY g.mm/m$^2$ day) |
|----------------|------------------------------|
| Nylon 11       | 18 – 25                      |
| HYTREL* 6356   | 15 – 20                      |

The measurements were made over a period of several months by accurate electronic weighing of sealed 2-metre tube lengths, including a metal reservoir, which had been pre-filled with methanol.

These values have been shown to increase by a factor of about 2 for every 10-15°C rise in temperature, but the effect of pressure is much less significant. However, as can be seen later, the use of improved hose liner materials can reduce these values by a factor of more than 100.


CORE TUBE MATERIALS

At this point it is appropriate to introduce some ideas for alternative hose liner materials and core tube constructions, which have been the subject of a major development programme over the last two years. It has now reached the stage where commercially produced hoses are under evaluation both in an extended Research Project as well as in actual service in an umbilical.

Bearing in mind the four major requirememts (chemical compatibility; permeability to methanol; mechanical performance; fluid cleanliness), as well as cost considerations, it became clear that a successful thermoplastic hose liner which optimised all these criteria may not be possible in a single material. In other words, a composite, or multi-layer tube was considered the best solution.

For example, the following distinct requirements were defined:

1.  A thin inner layer whose main purpose is to resist chemical attack.

2.  A layer (preferably thin) with outstanding barrier properties to resist fluid permeation.

3.  A "bulk" layer which would comprise the major proportion of the tube wall, and would provide the necessary long-term mechanical performance.

Of course, it would be an obvious bonus if layers 1 and 2 could be combined in a single material, thus simplifying the construction somewhat.  Other issues for a composite tube include the requirement for adequate adhesion between layers, ideally without the use of additional adhesive layers; the determination of "adequate adhesion" would have to be established by testing of actual hoses under extreme conditions of pressure and temperature cycling.  Typically the SAE J343 impulse test for high pressure hydraulic hoses carried out at 93°C and 133% of working pressure for up to 200,000 cycles.

## INNER LAYER – CHEMICAL DEFENCE

The requirement for this innermost layer is, above all, superior resistance to chemical attack by the wide range of fluids which – by design or otherwise – may flow through the hose during its lifetime.

One of the thermoplastic materials which has historically been considered to have good general resistance to a wide range of chemical is nylon (polyamide).  Nylon, in its various types, has been used in many fluid-contact application over the years, including, for example: fuel lines; automotive coolant systems; paint spray hoses; and of course hydraulic hoses – including offshore umbilicals, which also possess excellent toughness and overall mechanical performance.

As general purpose, chemically resistant materials nylons therefore offer valuable properties as inner "defence" layers in composite core-tube construction.  It may come as no surprise then that one of the multi-layer tube designs which has been developed makes use of a special nylon grade as its innermost layer.  This tube construction will be described in more detail later.

In the analysis of the functional requirements of this primary "defence" layer, however, it soon become clear that other materials, previously considered as either too exotic (in terms of cost) or lacking in certain other requirements as a general hose liner material, could now be realistically proposed for this inside surface layer.  Such thinking led inevitably to the fluoropolymer family of polymers.

Of the fluoropolymer materials, perhaps the best known is TEFLON*
PTFE - a material which has for many years been synonomous with
ultimate chemical resistance.  Typical industrial applications, for
example, are: lines for chemical plant, heat exchange tubes, exotic
fuel and solvent lines, etc.  Unfortunately, however, for use in long
length umbilical hoses, TELFLON* PTFE has one serious drawback - it
cannot be extruded continuously  into the long lengths required for
these hoses (this shortcoming is due to the fact that TEFLON* is
normally processed into tubes by sintering a compressed powder).  The
solution to the problem can, however, be found in one of the "first
cousins" of TEFLON* - ethylene tetrafluoroethylene copolymer (ETFE),
or TEFZEL*.

TEFZEL*, as its name implies, has many of the characteristics of
the TEFLON* resins, but unlike TEFLON* PTFE it is totally
thermoplastic material that can be processed and be converted by
conventional melt processing techniques.  It is a fluo-copolymer
polymer, and as such possesses similar inertness to a wide range of
fluids and chemicals, Table 2 below shows the basic mechanical
properties of TEFZEL*, whilst its overall excellent chemical
resistance can be judged from Table 3.

As an industrial fluorpolymer, TEFZEL* has been marketed since 1970,
and its applications are in areas such as chemical and pharmaceutical
plants.  Its credentials have been established for example as an
inner liner for automotive fuel lines, approved by manufacturers such
as GM and Chrysler in the US, where it also offers exceptionally low
permeability (this will be discussed later with regard to methanol
permeability).

In proposing TEFZEL* as the inner chemically-resistant layer in a
multi-layer core tube, it is of course essential to be able to
demonstrate sufficient adhesion to the outer layer(s) in order to
maintain a viable composite tube, able to withstand the rigours of
service as part of a high pressure hose  (pressure/temperature
cycling, etc).  It is, of course, well known that fluoropolymers are
particularly difficult to bond to other materials due to their inert
nature.  However, in cooperation with a major tubing manufacturer, a
technique to achieve such a bond by a unique and patented system has
now been developed and commercially proven.  This system, which will
be discussed again later, forms the basis for the second multi-layer
core tube construction which is the subject of this paper.


To summarise then, the inner-most surface of an umbilical hose core
tube must, as its primary function, resist chemical attack from a
wide range of chemicals and environmental conditions; it must be
extrudable continuously into long lengths; and it must retain good
adhesion to the remainder of the tube material.  The two tube
inner-layer materials which have been discussed above are believed to
offer unique properties in fulfilling these requirements.

## ANTI-PERMEATION (BARRIER) LAYER

The function of this layer is to contain the methanol and prevent migration into the outer layers of the core tube and thence to the remainder of the umbilical hose. It must also, of course, have no detrimental effect on the overall performance of the hose, and must therefore be suitably bonded to the other layers.

A material which has been widely used as a barrier layer, particularly in the packaging industry - as part of a multi-layer film or bottle construction - is ethylene vinyl alcohol, or EVOH. As such, the material is normally incorporated as a middle layer in a sandwich construction rather than having direct contact with the fluid to be contained.

EVOH can be extruded continuously in a coextrusion operation and most importantly it achieves a very strong bond with certain nylon types (specifically nylon 6 and nylon 6.12 types).

The combination of an EVOH barrier layer with a nylon inner and outer layer is therefore the basis for the first proposed core tube construction. In fact three different nylons have been tested in combination with EVOH - one is a semi-flexible nylon 6 based composition, known as ZYTEL* FE-7102, and the other is a more flexible elastomer mofified nylon 6 grade, ZYTEL* ST811. The third is ZYTEL* FE-3646, a plasticised 6.12 nylon. However, as will be discussed later, the first two are preferred for reasons of fluid cleanliness. The manufacturing process for the ZYTEL*/EVOH/ZYTEL* tube is essentially a standard tube coextrusion process, which is commercially proven in various tube applications. All calibration and other equipment downstream of the extruders is exactly as used today for mono-layer core tube production.

It should be pointed out at this stage that the combination of nylons with an EVOH barrier layer is the subject of an existing patent, and any use of this construction therefore requires an agreement with the patent holders. However, the fact that tubes with EVOH barrier layers are now sold in Europe for automotive fuel liners is an indication of the commercial viability of this process.

Test results obtained by measuring methanol permeability through actual core tubes having this construction are given later, however it is worth mentioning at this point that improvements of between 8 and 12 times compared with mono-layer (nylon 11 or HYTREL* tubes have been obtained.

The second major development in the area of barrier layers involves the use of fluoropolymers. It has been widely known for many years that materials such as TEFLON* and TEFZEL* offer extremely good performance in preventing permeation of fluids, including fuels and methanol. TEFLON* is in fact widely used for automotive fuel hose liners, for example under severe operating conditions, and particularly where very low permeation rates are specified by environmental legislation, such as in parts of the US, and more recently in Europe.

Bearing in mind the excellent performance of TEFZEL* as a chemically resistant liner, which was discussed above, it will be seen that there are obvious advantages in combining the chemical resistance properties of an inner TEFZEL* layer with its low permeability properties, and thereby achieving both desired functions from a single material. In conjunction with a nylon "bulk" layer - in order to provide overall mechanical performance and optimise the total cost of the tube - a simple 2-layer tube was developed, having been already in existance for some time as a patented construction for automotive fuel lines. This particular design was made possible by a proprietary system for bonding the inner TEFZEL* layer to the bulk nylon layer, which had been fully approved by US car companies as the best solution to the tough environmental legislation for low fuel permeation.

Work was therefore started on developing this TEFZEL*/nylon tube for application as a hose liner for offshore hoses. Prototype tubes were produced, by the patent holder, using a 0.3 mm inner layer of TEFZEL*, bonded to a ZYTEL* flexible nylon grade to construct a half inch core tube. These tubes were then tested for methanol permeability in the same manner as the ZYTEL*/EVOH/ZYTEL* tubes previously described. The results, which are discussed in more detail later, were extremely encouraging - improvements over nylon 11 and HYTREL* permeation rates by a factor of between 50 and 100 were obtained and reproduced in subsequent tests.

The next stage was to manufacture actual hoses in the two core tube constructions, viz: ZYTEL*/EVOH/ZYTEL*, and TEFZEL*/ZYTEL*. This was achieved through co-operation with one of the major producer of umbilical hoses and utilised normal technology of KEVLAR* braid and thermoplstic outer cover to manufacture half inch hoses for testing. These hoses have subsequently been put through extensive testing by the manufacturers, including high temperature impulse, tensile and crush tests, and of course permeability tests. The results confirmed the low permeability values obtained in the laboratory, and overall suitability of the core tubes for high pressure hose construction.

## PERMEABILITY TESTING

Permeability of fluids through polymeric materials is a complex process, and does not always follow expected rules. For example, as you would expect the permeation rate through a tube wall has been proved to be inversely proportional to wall thickness, in general – so doubling the wall thickness normally halves the permeation rate. However, in the case of multi-layer tubes, using mateirals of differing permeability properties, the situation is not so straight-forward. This has been demonstrated by varying the percentage thickness of certain barrier layers (such as a fluoropolymer) in relation to the overall tube wall thickness. In this case it has been shown that relatively thin barrier layers can achieve almost as good results as a thicker layer of the same barrier material.

Similarly, the effect of internal pressure has been studied, and the conclusion appears to be that for permeation of fluids such as methanol through polymeric materials having relatively low permeability coefficients, the effect of increasing pressure is relatively small. In other words, measurement of fluid permeability through unpressurised tubes will result in values which are of the same order of magnitude, and not widely different from a highly pressurised reinforced hose. On the other hand, the effect of temperature is of major importance. For relatively impermeable polymers, the temperature is believed to have an exponential effect on the rate of permeation. This implies that a small change in ambient temperature will result in a relatively large change in measured permeation values. Finally, the effect of the humidity surrounding the outside of the tubes under test will have an effect on measured permeability values – particularly for materials like nylon which can absorb relatively large quantities of water.

In the case of our multi-layer tube construction, therefore, it was decided to use the relatively simple test method specified in SAE J-30, but modified to use a 2 metre length of sample tube, plugged at one end, and attached to a small metal reservoir at the other end. Once filled with methanol, the tube plus reservoir were totally sealed and suspended vertically in a controlled environment. The loss of fluid from the tube was measured by a system of accurate electronic weighing of the complete system at regular intervals over a period of up to 3 months. Fluid lost through the tube was automatically replaced by that in the reservoir, so that the tube itself remained permanently full of methanol.

The results of this study are plotted in fig 1 below, calculated in terms of grams of fluid lost per day, per square metre of tube surface area, for 1 mm wall thickness. In calculating these results, the time to reach equilibrium was determined by establishing the point at which fluid loss became constant – a period of up to 3 weeks in the case of some tube constructions. It is interesting to put some of these results in perspective in terms of an actual hose carrying methanol in an umbilical: not allowing for the effects of outer braid and cover components in a hose, nor the pressure effects discussed above, the best results achieved in these tests represents a figure of approximately 9 grams of methanol lost per kilometre length for a half inch hose – at 20°C. At subsea temperatures of say 5°C (North Sea) the figure is further reduced to an almost insignificant value of less than 2 grams per kilometre per day.

## Average Permeation Rate (Methanol)
## gm.mm/m2.day

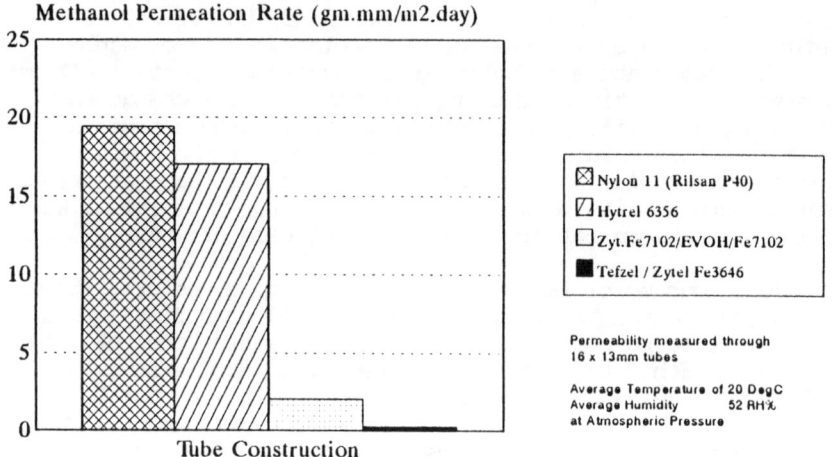

Figure 1

## CONTAMINATION EFFECTS ON THE FLUID

One of the most critical requirements for a hose liner used in
control fluid applications is that it must not contaminate the fluid
by extraction of solid or other components from the polymeric
material.  Although not so relevant for methanol injection lines, the
fact that hoses originally designed for methanol may potentially be
put to other use to carry control fluid, means that all hoses should
ideally be constructed to eliminate as far as possible the risk of
solid contaminants being extracted by control fluids from the polymer
lining.  As methonol is a particularly aggressive extraction agent,
it is sometimes used to test the "cleanliness" of hose liner
materials.

A study was therefore made of the residual methanol which had been
contained in the 2 metre permeation test tubes after a period of 3
months had elapsed.  This examination confirmed that the new hose
liners compare well with currently used materials such as Nylon 11
(RILSAN* P40 TLO) or HYTREL*; in fact the TEFZEL* lined composite
tube was by far the best in terms of detectable levels of extracted
contaminants.  This is indeed not surprising considering the
established credentials of fluoropolymers in terms of purity and
inert behaviour.


## FUTURE WORK

During the course of the development activities described above, it
has been the policy to work closely with hose manufacturers who have
contributed valuable input relating to performance requirements.
They have also greatly assisted, by the manufacture of test hoses and
by conducting in-house testing of hose samples.  These tests have led
to the positive results described above, however it is believed that
the most valuable data is likely to come from the independent R&D
program established by the industry-wide E&P Forum and being
coordinated through GEC Alsthom Ltd.  This study includes tests
conducted on strings of sample hoses held under controlled conditions
of temperature, and subject to cyclic pressure, over several months,
followed by examination and testing of the liner materials.
Permeability studies on the complete hoses are also included in this
programme.

By the time this paper is presented, it is likely that a commercial
umbilical containing hoses having a multi-layer construction as
described above will have been manufactured for sub-sea service.
However, other work is continuing to develop alternative multi-layer
constructions, using, for example, combinations of TEFZEL* and
HYTREL*.  Results to date are encouraging and again it is expected
that hose manufactures will be involved in evaluation at all stages
of the programme.

CONCLUSIONS

It is clear that the whole industry - from the oil companies through
the hose manufacturers to the raw material suppliers - now have a
much better understanding of the mechanisms and performance aspects
of hose liner design.  From the greater attention that has been paid
in recent years to understanding how failures can occur, and the
large amount of test data that has been generated, it is likely that
optimisation of all requirements may be better achieved by the use of
a multi-layer construction, rather than by a mono-tube design.
Taking into account these requirements, as well as cost considerations
and manufacturing feasibility, it is believed that 2 or 3 layer core
tube designs such as those described in this paper represent the
best way forward - not just for methanol lines,  but also for control
and chemical injection hoses.

Major programmes are ongoing to produce further improvements, making
use of the wide range of material combinations which are available in
order to achieve maximum effect in each of the critical areas of hose
performance.  It is believed that the use of multi-layer core-tubes
will herald a step-change in overall performance and reliability of
this important component of umbilical hose design.  With the
encouragement and commitment of all parties - oil companies, hose
manufacturers and raw material suppliers - it is hoped that the
advantages of thermoplastic umbilical hoses can be maintained long
into the future.

Note:      ZYTEL*, HYTREL*, TEFLON* TEFZEL* are all registered
           trademarks of Dupont.

           Rilsan* is the registered trademark of ATOCHEM SA

REFERENCES

LEBOVITS, A (1966) "Permeability ofpolymers to gases, vapors,and
liquids", Modern Plastics 43, March 1966, 139-150, 194-213.

LEBOVITS, A (1967) "The permeability and swelling of elastomers and
plastics at high hydrostatic pressure", Ocean Engineering Vol 1,
91-103.

HAXO, HE JR, MIEDEMA, J A & NELSON NA, MATRECON INC. USA,
"Permeability of polymeric membrane lining materials",

GOLDSBERRY, DR, CHILLOUS, SE & WILL RR, EI DuPont de Nemours and Co
Inc (1191) "Fluoropolymer Resins: Permeation of Automotive Fuels",
SAE Technical paper 910104.

| PROPERTY | TEST METHOD | UNITS | TEFZEL |
|---|---|---|---|
| Specific gravity | ASTM D792 | | 1,71 |
| Tensile strength | ASTM D638 | MPa | 44,8 |
| Ultimate elongation | ASTM D638 | % | 275 |
| Flexural modulus | ASTM D790 | MPa | 1 373 |
| Flex life | MIT (0,2mm, 180° flex) | | 30 000 |
| Impact strength | ASTM D256 +23°C ———————<br> -54°C J/m | | >1 100 |
| Hardness | ASTM D2240 | Shore D | 75 |
| Coefficient of dynamic friction | | | 0,4 |
| Melting point | | °C | 270 |
| Service temperature (20 000 h) with 50% retention of ultimate elongation | | °C | 155 |
| Inflammability | UL-94 | | 94 V-0 |
| Limiting Oxygen Index | ASTM 2863 | % | >30 |
| Heat of combustion | ASTM D240 | MJ/kg | 13,8 |
| Dielectric constant at $10^3$ - $10^6$ Hz | ASTM D150 | | 2,6 |
| Dissipation factor at $10^6$ Hz | ASTM D150 | | 0,005 |
| Arc resistance | ASTM D495 (Stainless steel electrodes) | s | 72 |
| Volume resistivity | ASTM D257 | $\Omega$ cm | $>10^{16}$ |
| Surface resistivity | ASTM D257 | $\Omega$ | $>10^{14}$ |

# Table II - General properties of TEFLON and TEFZEL

| CHEMICALS | TEMPERATURE °C | DAYS | % RETAINED TENSILE | PHYSICALS ELONGATION | % WEIGHT GAIN |
|---|---|---|---|---|---|
| **Organic chemicals** | | | | | |
| **Acids/Anhydrides** | | | | | |
| Acetic Acid (Glacial) | 118* | 7 | 82 | 80 | 3,4 |
| Acetic Anhydride | 139* | 7 | 100 | 100 | 0 |
| Trichloroacetic Acid | 100 | 7 | 90 | 70 | 0 |
| **Hydrocarbons** | | | | | |
| Mineral Oil | 180 | 7 | 90 | 60 | 0 |
| Naphtha | 100 | 7 | 100 | 100 | 0,5 |
| **Aromatics** | | | | | |
| Benzene | 80* | 7 | 100 | 100 | 0 |
| O-Cresol | 180 | 7 | 100 | 100 | 0 |
| **Amines** | | | | | |
| Aniline | 120 | 7 | 81 | 99 | 2,7 |
| n-Butylamine | 78* | 7 | 71 | 73 | 4,4 |
| Di-n-Butylamine | 120 | 7 | 81 | 96 | N.D** |
| Pyridine | 116* | 7 | 100 | 100 | 1,5 |
| **Chlorinated Solvents** | | | | | |
| Carbon Tetrachloride | 78* | 7 | 90 | 80 | 4,5 |
| Chloroform | 61* | 7 | 85 | 100 | 4,0 |
| Dichloroethylene | 32 | 7 | 95 | 100 | 2,8 |
| Methylene Chloride | 40* | 7 | 85 | 85 | 0 |
| **Ether** | | | | | |
| Tetrahydrofuran | 66* | 7 | 86 | 93 | 3,5 |
| **Ketones** | | | | | |
| Acetone | 56* | 7 | 80 | 83 | 4,1 |
| Acetophenone | 180 | 7 | 80 | 80 | 1,5 |
| Cyclohexanone | 156* | 7 | 90 | 85 | 0 |
| Methyl Ethyl Ketone | 80* | 7 | 100 | 100 | 0 |
| **Esters** | | | | | |
| n-Butyl Acetate | 127* | 7 | 80 | 60 | 0 |
| Ethyl Acetate | 77* | 7 | 85 | 60 | 0 |
| **Polymer Solvents** | | | | | |
| Dimethylformamide | 90 | 7 | 100 | 100 | 1,5 |
| Dimethylformamide | 120 | 7 | 76 | 92 | 5,5 |
| Dimethylsulfoxide | 90 | 7 | 95 | 90 | 1,5 |
| **Inorganic Acids** | | | | | |
| Hydrochloric (Conc.) | 23 | 7 | 100 | 90 | 0 |
| Hydrochloric (Conc.) | 106 * | 7 | 96 | 100 | 0,1 |
| Hydrobromic (Conc.) | 125 * | 7 | 100 | 100 | N.D** |
| Hydrofluoric (Conc.) | 23 | 7 | 97 | 95 | 0,1 |
| Sulfuric (Conc.) | 100 | 7 | 100 | 100 | 0 |
| Sulfuric (Conc.) | 120 | 7 | 98 | 95 | 0 |
| Nitric - 25% | 100 * | 14 | 100 | 100 | N.D** |
| Nitric - 50% | 105 * | 14 | 87 | 81 | N.D** |
| Nitric - 70% (Conc.) | 23 | 105 | 100 | 100 | 0,5 |
| **Inorganic Bases** | | | | | |
| Ammonium Hydroxide | 66 | 7 | 97 | 97 | 0 |
| Potassium Hydroxide (20%) | 100 | 7 | 100 | 100 | 0 |
| Sodium Hydroxide (50%) | 120 | 7 | 94 | 80 | 0,2 |
| **Peroxide** | | | | | |
| Hydrogen Peroxide - 30% | 23 | 7 | 99 | 98 | 0 |

## Table III - Effect of various chemicals on TEFZEL

# Session 4
# Application Engineering

# SUBSEA WELL CONTROL SYSTEMS
# THE SPECIFICATION OF
# RELIABILITY, AVAILABILITY AND MAINTAINABILITY

STEVE BYRNE
Total Oil Marine Plc.
Stell Road
Aberdeen AB1 2QR

## ABSTRACT

The successful design of a subsea well control system depends chiefly upon the selection of highly reliable subsea components. If a system fails regularly, due to unreliable components, then the cost to maintain that system may quickly escalate and surpass capital cost.

Poor specification means that vendors are often constrained into tendering their cheapest solution, in order to maintain competitiveness. The cheapest solution may appear to be 'fit for purpose', but for how long?

It is not sufficient merely to specify the reliability or availability of a system, there are differing requirements for the topsides and the subsea elements. It is well known that reliability data for electronic components exists, but for other components, such as subsea mateable connectors, not enough reliable data exists. For these components additional criteria should be specified to ensure that reliability is addressed during selection.

It is traditional for a reliability study to be carried out during the detailed design phase. A reliability study alone is not sufficient, it may be 'closing the gate after the horse has bolted'! The quantitative specification of reliability, availability and maintainability, before contract award, is the key to ensuring that the system design will be adequate.

*Volume 32: Subsea Control and Data Acquisition,* 187–204.
© 1994 *Society for Underwater Technology. Printed in the Netherlands.*

## INTRODUCTION

This paper sets out the approach to a quantitative specification of reliability. It seeks to illustrate that a realistic specification is possible and it introduces a yardstick by which any subsea well control system may be specified. For simplicity, a single subsea well control system, controlling an oil producer, is considered. The electro/hydraulic umbilical is not included, as it is assumed that it would be specified separately.

Reliability has different meanings to different individuals. Some think of reliability in a non-quantifiable way, much the same as dependability, and some confuse it with availability. Reliability can be very specific, yet rarely is it quantitatively specified. It is more common to see availability specified, even though, as we shall see later, availability is a rather non-specific term.

The reliability of subsea control components, particularly electronic components, has been the subject of many papers and studies but there is little in the way of guidance for the Operators. How does the industry want us to specify systems so that vendors can be confident that they are bidding competitively? The Operators, for their part, want a quality product in order to avoid extensive repairs in the future. We must state therefore, exactly what we need, particularly in terms of a system's ability to operate successfully throughout the life of a development.

The prevailing view amongst many Operators is that the industry has reached a suitable level of maturity and that we need to 'cut our cloth' to suit. Advances in electronics, distributed sensors and connectors prove that we can further enhance reliability with little increase in cost. We should all assist the industry in trying to achieve higher reliability. Opting for low cost, low reliability solutions, in critical areas, puts the Operator at risk and is blatant 'short termism'.

A successful design and a cost effective solution can only be achieved with painstaking attention to reliability throughout the design process. The design process involves:

- Thorough analysis of operability and maintainability;
- Detailed and quantitative reliability specification;
- Thorough follow-up analyses to ensure that the level of reliability is achieved;
- Commitment to improving reliability where possible and where cost effective.

## SUBSEA WELL CONTROLS

A subsea well can be controlled in a number of different ways depending on the distance of the wellhead from the production facility and whether there is a requirement for sophisticated monitoring.

Wellhead valves are controlled by any one, or a combination, of the following methods:

- Direct Hydraulic;
- Piloted Hydraulic;
- Sequential Piloted Hydraulic;
- Multiplexed Electro-Hydraulic;
- Acoustic Electro-Hydraulic.

If production quotas are important, then some form of back-up is normally employed. The type of back-up is also influenced by the distance of the wellhead from the production facility.

Fig. 1 below details the subsea elements of a typical control system:

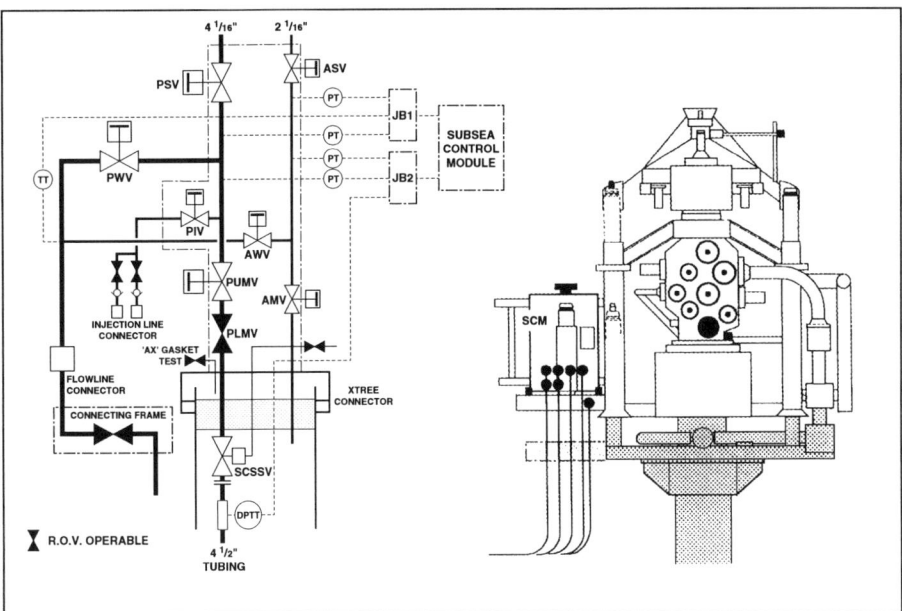

**Fig. 1  Subsea Well Control Schematic and Wellhead General Arrangement**

The subsea well control system comprises a topsides and a subsea element. The configuration of both elements is best determined by an operability and maintainability study, in conjunction with a failure mode effects and criticality analysis (FMECA).

Fig. 2 below details the main topsides and subsea elements of a simple non-redundant control system:

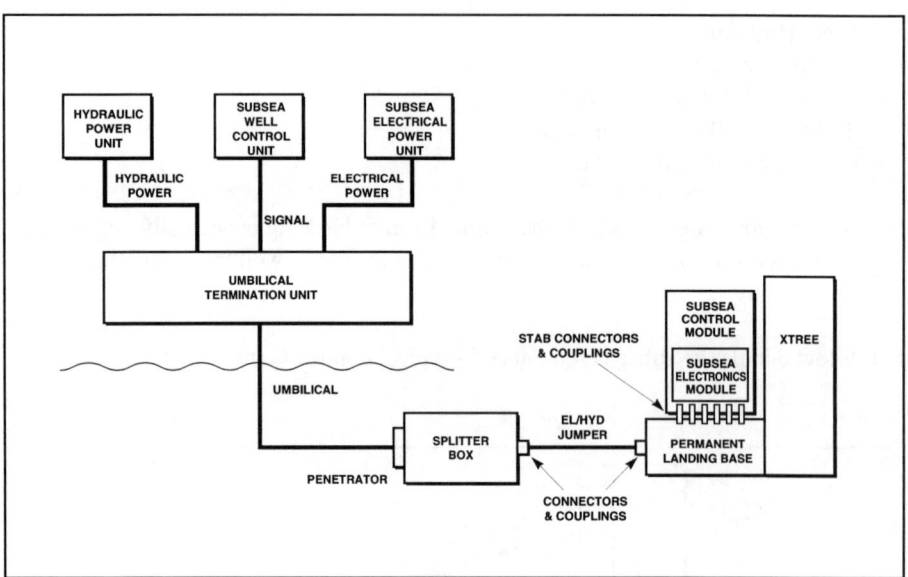

**Fig. 2    Block Diagram of a Simple Subsea Well Control System**

Before discussing the methodology for arriving at the final configuration, it is worth looking at what is meant by the terms reliability, maintainability and availability.

## RELIABILITY, AVAILABILITY AND MAINTAINABILITY

### Reliability

A hardware reliability analysis is based upon statistical fundamentals whereby the probability of a system operating without failure is defined as its 'reliability' or more succinctly its 'probability of survival'.

The following concepts help to relate reliability and failure:

- The probability of a given system failure occurring is a combination of the individual module or component failure probabilities;
- The failure rate (number of failures per unit time) typically follows the bath-tub curve as shown in Fig 3;
- Three distinct failure zones exist throughout a product's life. They are the infant mortality zone, the random failure zone and the wear-out zone. They are distinguished by failure frequency.

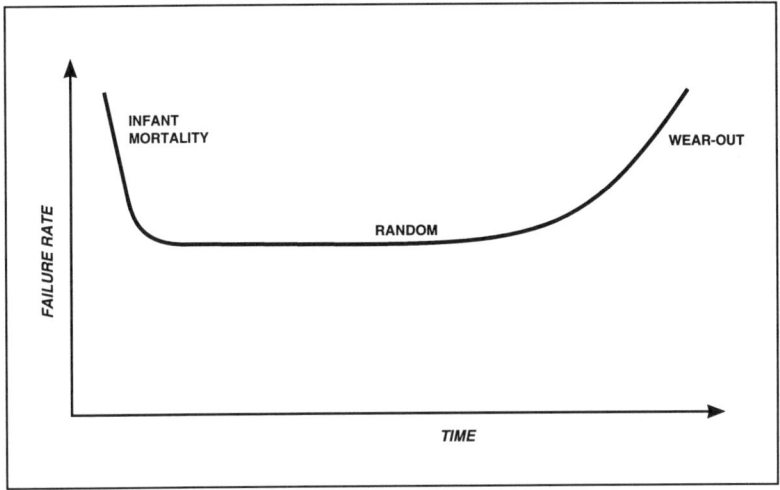

**Fig. 3  Bath-Tub Failure Rate Curve**

Infant mortality is early life failure and it is usually associated with material or manufacturing defects. When the wear-out failure is reached the failure rate increases rapidly and the end of the useful life of the component is reached.

For control systems, we are usually only interested in the region of time independent failure rate caused by random events. Time independent failures are exponentially distributed and such failures predominate in electronic devices.

Complex failures, such as failures associated with dynamic parts, may be modelled by a distribution which is used to describe all three regions of the bath-tub curve, the Weibull distribution.

The probability of survival (reliability), for a specific period of time, is related to the failure rate by an exponential expression:

$$R = e^{-\lambda t}$$

where:

R   =   reliability;
$\lambda$   =   failure rate (failures/time);
t   =   specific time period.

Where there are a number of components in series:

Rs   =   Ra.Rb.Rc.....etc.

where:

Rs   =   system reliability;
Ra, Rb, Rc   = component reliability.

Where there are a number of components in parallel:

Rs   =   1-(1-Ra)(1-Rb)(1-Rc).....etc.

Mean time between failure (MTBF) is the inverse of the failure rate:

$$MTBF = \frac{1}{\lambda}$$

and $R = e^{-t/MTBF}$

The 1 year reliability of a system with an MTBF of 5 years is 82%. Increasing the MTBF to 10 years gives a reliability of 90%. To increase the reliability to 99%, an MTBF of 100 years is required!

**Availability**

Availability, rather than reliability, is often specified.

$$\text{Availability (A)} = \frac{\text{Up-time}}{\text{Up-time} + \text{Down-time}}$$

or
$$A = \frac{\text{MTBF}}{\text{MTBF} + \text{MTTR}}$$

where:

      A   = Availability;
   MTBF = The statistical mean time between failures;
   MTTR = Mean time to repair.

**Maintainability**

Maintainability is an area where little or no analysis is carried out. Designers rarely inquire into maintenance philosophy, yet maintenance considerations can significantly affect the configuration of a system. Software maintenance, in particular, receives little attention.

In hardware systems, maintainability has a direct influence on the reliability of systems that have redundancy, and upon the availability of all systems. Maintainability analysis and specification should therefore be regarded as a necessary adjunct to reliability specification. Maintainability analysis may be used to determine servicing, condition monitoring and repair down-time predictions.

Before we move onto reliability specification, we should look at the correlation between reliability, availability and mean time between failure (MTBF).

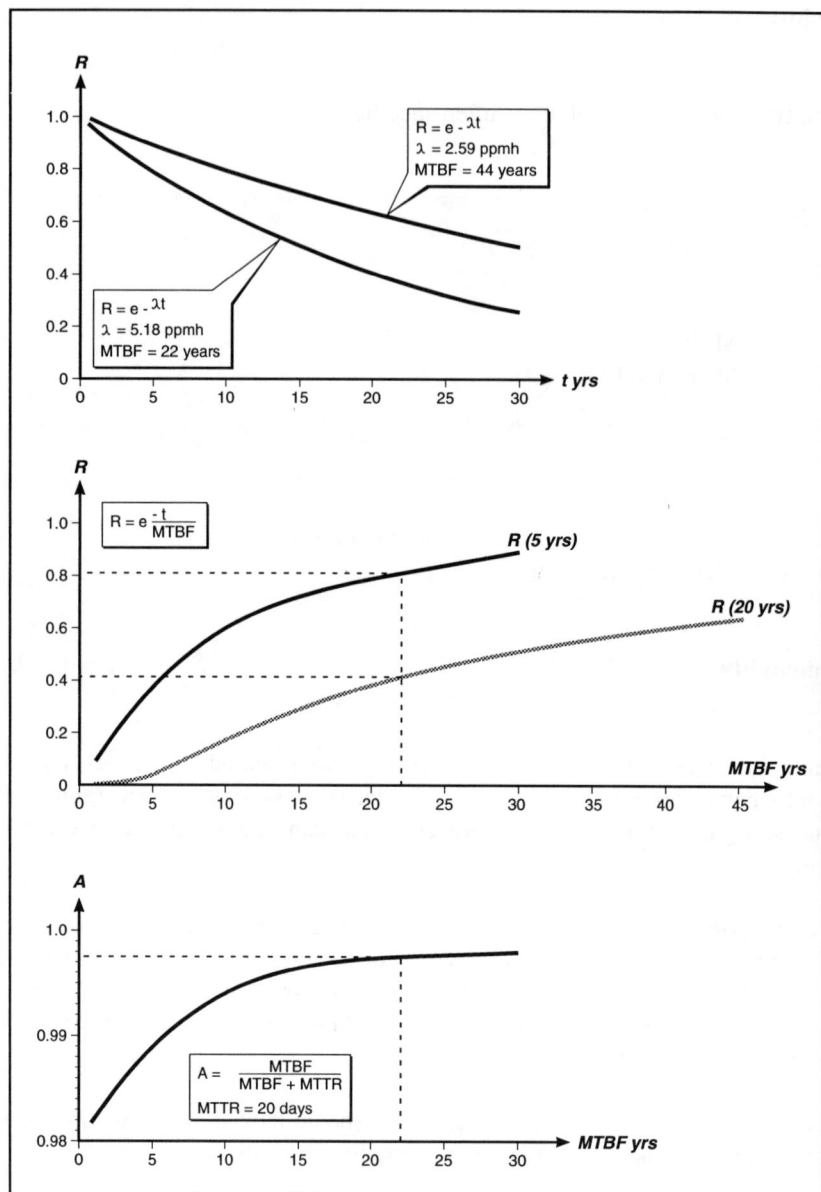

**Fig. 4   Reliability or Availability**

Reliability curves are exponential, the law of 'diminishing marginal returns' could be said to apply. There comes a point, where a further increase in reliability is not cost effective.

Fig. 5 below illustrates the point.

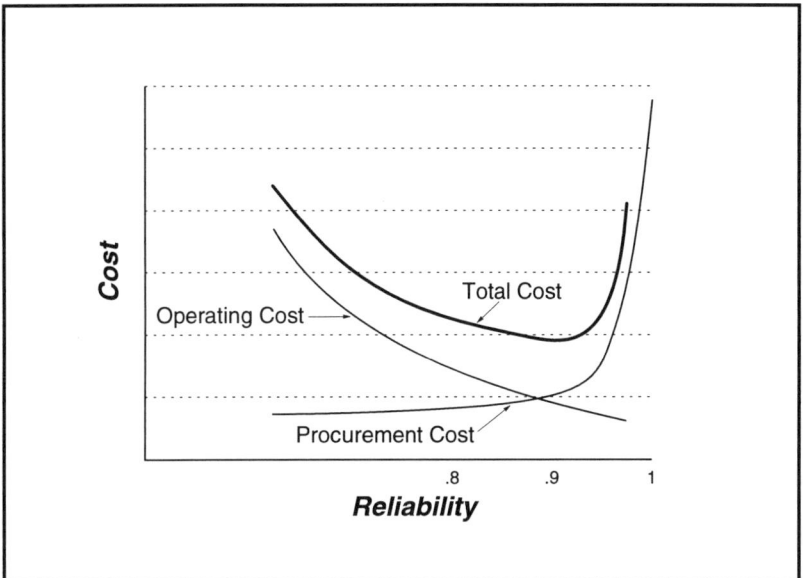

**Fig. 5 Cost v Reliability**

As most critical components in a subsea well control system have MTBFs of much greater than 30 years, then it is reasonable to assume that a 5 year system reliability of between 80% and 90% is achievable. A 5 year reliability is usually specified, as it is expected that maintenance intervention will be carried out at least once every 5 years and perhaps even more frequently.

Curiously, it can be seen from Fig. 4 that, increasing a system MTBF above 22 years does not present any significant increase in availability. It follows therefore, that if system reliabilities of between 80% and 90% are achievable, then the specification of availability for a subsea control system becomes pointless, particularly considering that the MTTR is very difficult to predict.

## SYSTEM CONFIGURATION

Before a reliability specification can be completed, it is necessary to know the configuration of the system, as obviously a dual redundant system will be more reliable than a non-redundant system. Redundancy, if employed, may be standby or active, uniform or diverse. The systematic evaluation of the operability and maintainability requirements, failure modes, failure conditions and failure effects should help us in this direction.

Cost, it must be remembered, is also part of the equation. An analysis of CAPEX (Capital Expenditure) versus OPEX (Operating Expenditure), based on the desired configuration, should always be used to confirm that the right decisions have been taken. The long term maintenance costs, over the lifetime of a development, can be very significant. The cost of intervention, by a diving support vessel (DSV), to recover a Subsea Control Module (SCM) can be of the order of £350,000 at 1993 prices. The capital cost of a redundant Subsea Control Module might be of the order of £300,000, or £200,000 for a non-redundant version.

### Operability & Maintainability

- Is the subsea system to be diver or diverless maintained?
- Is production required to be maintained? If not, what is the maximum shutdown time that can be tolerated?
- Is there a requirement to maintain any part of the system whilst maintaining production?
- Does the system carry out any ESD (emergency shutdown) functions?
- What level of system diagnostics is required?
- Is the system software fully developed in both single or dual redundant modes?
- Is a software maintenance programme in place?

### Use of FMECA

- It is to identify potential weaknesses;
- It is normally initiated early in a programme;
- Individual plans will be required to allow for variations in system complexity.

**Failure Modes**

- Loss of electrical power;
- Loss of hydraulic power;
- Loss of communications;
- Failure of a component or components to operate within a prescribed time;
- Failure of a component or components to cease operation within a prescribed time;
- Failure of a component or components during system operation.

**Failure Conditions**

- No single point of failure should inhibit the shutdown of a well;
- No single point of failure should prevent the operating personnel from assessing and evaluating system status;
- No single point of failure should prevent operating personnel from diagnosing and rectifying the failure (within a time period, dependent upon the importance of production);
- No single point of failure should induce failures into a redundant part of the system;
- Loss of electrical power should cause the wellhead to shut in;
- Loss of hydraulic power should cause the wellhead to shut in;
- Loss of communications should cause the wellhead to shut in.

**Failure Effects**

- Complete loss of production;
- Partial loss of production;
- No loss of production;
- Complete loss of monitoring;
- Partial loss of monitoring;
- No loss of monitoring;
- System fails safe;
- System fails to hazard.

## THE SPECIFICATION OF RELIABILITY

Having determined the configuration of the system, reliability can now be specified.

Functional design specifications often contain statements such as:

- The system shall be inherently reliable, or the system shall be backed-up to ensure a verifiable up-time of 99.5%.

Impressive words, but they are actually meaningless! Availability, for example, cannot be verified unless an MTTR is specified.

Traditionally, reliability has been specified qualitatively rather than quantitatively. Where reliability is not paramount, this approach may be perfectly acceptable. For subsea well control systems, however, where reliability is crucial, a quantitative specification is needed.

A quantitative reliability specification should include the following clauses:

- The criteria for failure;
- The reliability characteristics that are appropriate (e.g. MTBF, MTTR etc.);
- The required value of the reliability characteristics;
- The time during which, and the conditions in which, the system is required to perform its function.

The clauses should also take into account the following:

- Environmental conditions;
- Stress conditions;
- Common mode failures;
- Design life;
- Maintenance;
- Storage life;
- Manufacturer's reliability testing;
- Customer testing.

Topsides equipment is different to equipment located subsea in terms of its reliability requirement. It is of paramount importance that the subsea equipment is reliable, but maintainability is the main criterion for the topsides equipment. The reliability specification should therefore be written in two main parts.

**The Subsea Elements**

Refer to Fig. 2 - Block Diagram of a Simple Subsea Control System.

In order to ensure that reliability can be specified, it is useful to know what might typically be expected. Vendors and manufacturers are usually quite willing to divulge failure rate data for individual components. The simplified Reliability Block Diagram (RBD), as shown in Fig. 6 overleaf, indicates the different levels of reliability between typical redundant and non-redundant systems.

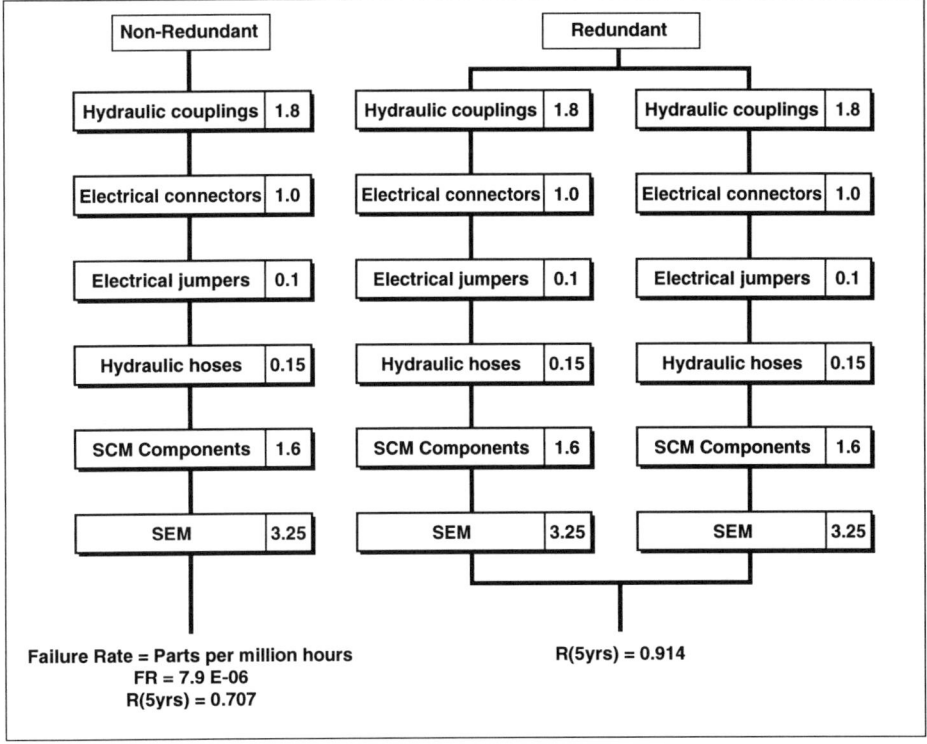

**Fig. 6  Simplified Subsea Controls Reliability Block Diagram**

The subsea system should be broken down into elements and each element should be addressed separately.  Critical components should be considered in light of their effect on the rest of the system, or the effect of the rest of the system on them.  The main elements might be as follows:

Critical:

- Subsea Electronics Module (SEM);
- Directional control valves;
- Pin connectors;
- Hydraulic couplings;
- Hoses;
- Cables.

Non critical:

- Pressure sensors;
- Temperature sensors.

If a 5 year subsea system reliability of over 80% is required, then the MTBFs of each critical component must be high enough to ensure that when all the series failure rates are added, the resultant MTBF is greater than 22 years.


## Subsea Electronics Module (SEM)

Vendors are now confident that they can achieve a SEM MTBF of well over 30 years. When specifying SEM reliability, environmental, quality and stress factors, as well as operating temperature should be stated. If a full specification is given, then we can be confident that the vendors will be 'playing on the same level playing field', when bidding systems. The reliability predictions that are included in the bids may then be comparatively assessed.

A typical specification might contain the following statements:

Failure rates are to be calculated in accordance with Mil-Hdbk-217F;
SEM failure rate is to be calculated on a 'parts count' basis, all parts are to be included;
SEM  MTBF > 35 years;
Environmental factor - Ground Fixed;
Stress factor $\leq 0.5$;
Temperature + 25 °C.

Quality Levels:

- Microcircuits - S;
- Discrete Semiconductors - JANTXV;
- Capacitors, Established Reliability - D;
- Resistors, Established Reliability - S;
- Coils, Moulded, R.F., Established Reliability - S;
- Relays, Established Reliability - R;
- Non-Established Reliability Parts - Mil-Spec.

Specifying reliability levels helps to ensure a reliable product, however, a client will have greater confidence if he receives evidence of high MTBFs. It is therefore desirable that vendors should be instructed to return calculated and historical figures in their bids.

## Non-Electronic Components

It is relatively simple to specify a minimum MTBF for each of the non-electronic components as data can be obtained from vendors and manufacturers. However, to be sure of quality, materials and performance criteria should also be specified. Typically the reliability specification should include:

Design life qualification;
Minimum insulation resistance values, at specific voltages, for cables and connectors;
Minimum pitting resistance equivalent number (PREN) for connector/coupling metal parts exposed to seawater;
Minimum number of sealing levels on the front and back ends of connectors;
Qualification of compatibility of materials, separately or together, in seawater or control fluid.

## The Topsides Elements

Refer to Fig. 2 - Block Diagram of a Simple Subsea Control System.

Although topsides equipment failure is just as likely to bring a system down as a subsea element failure, maintainability and redundancy considerations are more important than reliability. The final system configuration will have been arrived at previously, so any additional redundancy, which may be required to allow on-line maintenance, should be specified for. Again, elements can be split out separately and identified as critical or non-critical.

Critical:

- Subsea Well Control Unit;
- Subsea Electrical Power Unit;
- Hydraulic Power Unit;
- Umbilical Termination Unit.

Non-critical:

- Printer.

Each individual element should be studied for maintainability. If maintenance has to be carried out, whilst the system is on-line, then units should be designed to allow for this.

## Subsea Well Control Unit

The Subsea Well Control Unit should be designed for ease of maintenance, all components should be easily accessible. Printed circuit cards should be front-mounted for ease of removal. The removal of any individual component should not require the removal of other components or the dismantling of the panel. For maintainability reasons alone, it is always useful to have dual redundant controllers. Dual controllers allow for the extensive use of diagnostic facilities in fault finding. Reliability or MTBF for the controllers should be specified. In the absence of a specification, vendors would probably offer non-industrial grade equipment.

## Subsea Electrical Power Unit

The Subsea Electrical Power Unit should be designed in a similar fashion to the Subsea Well Control Unit. It should be possible to isolate channels for maintenance whilst maintaining other channels live. All components should be accessible for maintenance.

## Hydraulic Power Unit

The maintainability study will have identified certain requirements. If on-line maintenance is required, then components relating to an individual function should be separated from others. Block mounted solenoid valves are not a good idea if the removal of one requires the isolation of others. It should be possible to temporarily bypass individual functions whilst maintenance is carried out. Diagnostic facilities such as check valve/bleed valve arrangements and gauges should be incorporated to allow fault diagnosis down to component level. All components should be easily accessible and removable. All wetted parts and parts exposed to saliferous atmospheres should be manufactured from suitable, corrosion resistant materials.

## Umbilical Termination Unit

The unit should be designed with maintenance in mind. Components must be accessible for removal and hydraulic lines downstream of block/bleed valves should have gauges installed.

## DESIGN PROCESS

Having identified the studies and specifications that are part of the design process, it is necessary to obtain the right information and to carry out the necessary studies at the appropriate time.

Fig. 7 is a typical design process flowchart that includes a reliability programme.

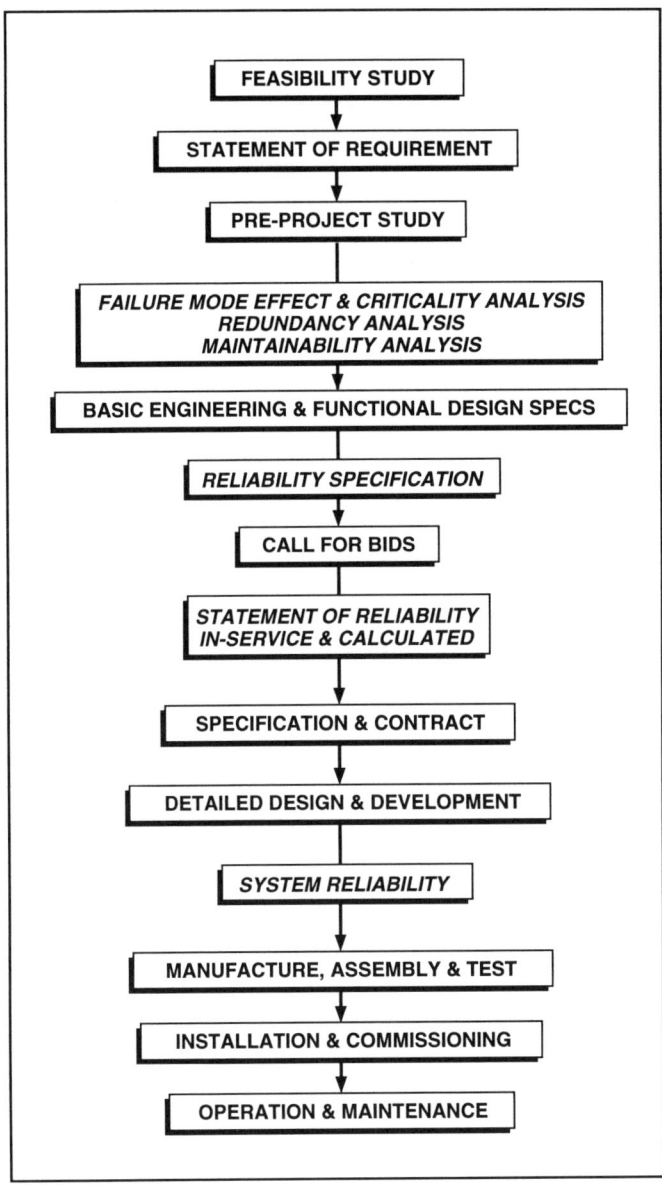

**Fig. 7 Subsea Well Control System Design Process**

## CONCLUSIONS

- A thorough, quantitative reliability specification is required to ensure that vendors can bid on a 'like for like' basis.
- Experience has shown that scope exists for further improvement in component reliability, with a commensurate improvement in system reliability.
- For a subsea control system, with an expected MTBF of greater than 22 years, the consideration of availability, in mathematical terms, is inappropriate.
- A 5 year reliability of 90% for the subsea element of a subsea well control system, excluding the umbilical and non-critical components, is achievable.
- A reliability programme, early in the design process, is necessary to ensure a successful and cost effective solution.

## REFERENCES

1.  Schrufer, E. (1986) Introduction to Reliability Modelling. Proceedings of an IFAC Conference, Reliability of Instrumentation Systems, The Hague, Netherlands.

2.  Renner, K.M. (1988) Availability, Reliability & Redundancy....Relieving the Confusion. Proceedings of an IFAC Worshop on Industrial Process Control Systems - Reliability, Availability and Maintainability, Bruges, Belgium.

3.  Goble, W.M. (1992) Evaluating Control Systems Reliability, Techniques and Applications, Instrument Society of America, North Carolina, U.S.A.

# THE APPLICATION OF PROPRIETY DCS/SCADA EQUIPMENT FOR SUBSEA MONITORING AND CONTROL SYSTEMS INCLUDING SMART SENSOR INTERFACE - A SYSTEMS OVERVIEW

Eur. Ing. G.R. CREECH Bsc MIEE M Inst. MC and P.L. SMITH
Hasbury Communications Limited
Paragon House, Ivanhurst Industrial Estate,
Woodham Road, Battlesbridge, Essex, SS11 7QL
England.

## ABSTRACT

The enhanced reliability and down sizing of electronic equipment in general, together with the increasing power of microprocessors and the use of Industry Standard Protocols, has now opened up more than ever before the opportunity for Subsea Instrumentation Systems, to utilise Standard Instrument and Control equipment.

## INTRODUCTION

The current exploitation of marginal oil and gas fields has resulted in the use of more cost effective methods of production, such as subsea wellheads. Control and Instrumentation just as in any other oil and gas installation, has a major role to play in the success of such applications. However, what should be a straight forward and simple application of a small DCS/SCADA System is often complicated and over engineered to compensate for the perceived 'harsh environment' into which the equipment is being placed, especially that which is being located subsea. In actual fact, there are many propriety elements of such a DCS/SCADA System available today which, due to the state of current technology and Quality Assurance, are together suitable for such applications without recourse to over specified and therefore excessively expensive solutions. Whilst providing due consideration to the problems of Subsea/Topsides monitoring and control applications, the use of P.C. workstations, P.C. based process control software packages, dual high security communications and remote intelligent I/O, can provide for a highly featured, secure and cost effective system.

Volume 32: Subsea Control and Data Acquisition, 205–210.

Moreover, the established use of Smart transmitters which communicate digitally has meant that the 'reading' of Process values can now be made at the topsides terminal with a much greater degree of accuracy than via a conventional Analogue card. Also possible with Smart sensors are enhanced self diagnostics and the ability to remotely re-range and re-calibrate the transmitters from the topsides terminal. This paper will examine just such a system and provide an overview of what can be achieved simply and effectively.

## EQUIPMENT OVERVIEW

In considering any Subsea Monitoring and Control System, there are three main elements; the Subsea equipment, the Communication equipment and the Top sides Monitoring and Control Sub-system. We shall take each of the above in turn and briefly examine proprietary equipment which can be used in the Subsea environment, with little or no changes other than application specification engineering.

## SUBSEA INTERFACE EQUIPMENT

This category of equipment can be broken down into three sub-headings; Primary Transducer, Electronic Interface and Communications.

Taking each of the above in turn we will look at what is currently commercially available, as follows:

### Primary Transducer - Conventional

There is currently available a multitude of conventional transducers suitable for installation in some extremely harsh environments - far more onerous than can be found in the relatively stable environment of a Subsea instrument housing. Such transducers can be chosen, not only for their particular interface characteristics, i.e. pressure, temperature, flow, but also chosen for their reliability, quality and construction medium i.e. stainless steel, flameproof, etc. Such transducers provide industry standard outputs such as 4-20mA, voltage and milli-volt signals which can be "read" by virtually all I/O proprietary input/output equipment.

## Primary Transducer - Smart

Of increasing popularity within the industrial sector are Smart Transmitters, which have the ability to communicate to interface equipment, information regarding the transmitter status other than the standard primary variable. This is done by the use of imposing a digital signal on the cables conventionally used to carry the 4-20mA signal. Honeywell for example, using their DE Transmitters, can measure the primary variable and communicate to the interface equipment either in conventional format or in digital format, but not simultaneously. The choice has to be made upon installation which method is to be utilised.

The Hart protocol on the other hand, enables the multi-dropping of transmitters in digital mode or, if only one device is connected on to the loop, the transmitter head can be used simultaneously to provide conventional 4-20mA signal and digital information. The digital communication is bi-directional and therefore, in addition to being able to integrate the transmitter for primary and secondary variables, other information can be requested, including transmitter type, manufacturer, date of installation, date of last calibration and other propriety information, which the transmitter manufacturer had built in to the device using the Hart data base. Of particular importance for Subsea systems, the application of Smart Transmitters should be considered for two main reasons. Firstly, if used in digital mode only, no Analogue to Digital (A-D) conversion is used and therefore the accuracy of the primary values tend to be much higher than when using conventional A-D loops. Secondly, due to the very nature of the application, the devices can be re-ranged and calibrated manually and this is an advantage that the Smart Transmitters have, in as much that they can be accessed digitally from the Top-side computer or even from an onshore maintenance terminal and down loaded with the new information automatically.

A print out of the transducers current set-up information can then be printed and filed, therefore complying with ISO9000 requirements. Again, there are a multitude of transmitter types and manufacturers to choose from with a correspondingly wide choice of housings and environmental specifications from which to choose. In particular, it should be noted that Smart Transmitters are being used exclusively on new unmanned platforms within the North Sea for the very reasons given above.

## Electronic Interface

When considering the type of electronic interface required for Subsea Instrumentation systems, two factors bear heavily on the type of equipment used. These two factors are; whether the interface equipment will be required to carry out any local control or whether the interface is being used merely as a multiplexor device. Whichever option is chosen, ideally the equipment should have the ability to easily integrate Smart Transmitters, contain integrally the communications equipment/interface required, with the ability to be used in either an intelligent, i.e. programmable or monitoring/multiplexing format only.

The equipment should also be capable of being used in a redundant format for higher availability and for 'n' out of 'n' voting if used for local control. There are many Input/Output (I/O) sub-systems which can be used for this application and they range from DIN Rail mounted I/O devices to programmable logic controllers right through to small DCS systems. However, current technology and the use of surface mount techniques coupled with greater quality assurance awareness, has meant an overall increase in the reliability of electronics in general and field interface equipment is no different. It is therefore possible to find a multitude of suppliers of I/O sub-systems, whose equipment requires no specialised engineering apart from application oriented engineering which are entirely suitable for Subsea applications. For example, it is possible to use an off-the-shelf I/O device such as that shown in this diagram, which is small, compact, low power, has inherent logic capability, can be used with Smart Transmitters, has an on board watchdog, uses dual independent communication links , can be multi-dropped to other I/O devices, uses a de-facto industry standard protocol and has full CRC 16 error checking on its communication links and yet still costs less than £1000! This device can also be used for local control where the control algorithm is down loaded via the communication link from a top sides computer; (PC or equivalent) and the device used in a two out of three voting configuration. Other software configuration tools that can be used to provide higher integrity, are the use of 'highest' or 'lowest' analogue readings of 'n' number of analogue readings.

## Communications

Typical Subsea control systems are connected to the surface vessel via an umbilical cord, which provides both communications and power to the Subsea instrumentation system. In all cases, dual communication links should be used as a minimum to provide high availability and if possible the interface equipment should have the capability to utilise both links simultaneously and not be a pseudo dual link i.e. dual communication links connected to only one serial port on the interface equipment with a watchdog change over switch. Typically in the North Sea, the umbilical cable tends to be between one to three kilometres in length and care and consideration should be given to the screening of the cable for isolation against electrical noise induced from power cables running along side. The integrity of each link should be increased by the use of Cyclic Redundancy Checking (CRC 16 bit error checking).

In this particular application shown in this diagram, dual RS485 communications were employed on an umbilical cord comms link, almost two kilometres in length and the system performs well. Besides the problems of noise on the power cable running alongside the serial communications link to the sea bed, problems with the communications link can be induced from Top-side radar signals that occur on the cable when it reaches the ship. This is usually exposed and therefore consideration must be made as to the routing of the communications cable and its Top-side screening to prevent any such induced signal noise.

## TOP SIDES MONITORING AND CONTROL SYSTEM

The choice of equipment used to monitor and control the Subsea instrumentation system can range from a single workstation VDU terminal through to a full blown DCS system depending upon the size of the system and the cost budget available. There are many software packages today which run on PC work stations which challenge many of the DCS vendors Man Machine Interface (MMI) and control packages in terms of functionality, ease of use and certainly, cost. It is interesting to note that many DCS companies are now offering PC based solutions for the smaller systems, which is a substantial move from the stance they were taking, just a few years ago.

It has to be recognised that the power of the new PC's and their development programme of increasing power over the next few years is beginning to make the PC work station a completely viable and attractive means of offering a Man Machine Interface for many systems and in terms of an overall view of the system it is probably true or at least becoming true, that in many instances the only advantage that a DCS company has over a systems integration vendor offering propriety solutions, is the DCS company's traditional application knowledge. This however, is becoming eroded rapidly by smaller integrators gaining experience through the use of smaller PC based systems.

A modern PC based software package such as Intec's Paragon TNT for example, not only enables 'Real Time' plant monitoring, using graphical representations of the system, but can also be easily configured by a non-software engineer. It also has a library of common protocol drivers and has very powerful third party software links enabling dynamic data exchange between many of the common PC based third party software packages, such as Lotus 123, Excel, etc. It also has Structured Query Language (SQL) links to more powerful Unix based software data bases such as Oracle, etc. Software packages such as Paragon TNT are revolutionising the instrumentation and control market and used together with propriety I/O systems, a very powerful high availability system can be bought together without the need for specialist engineering and more importantly a cost penalty for using such equipment in Subsea applications.

## CONCLUSION

The environment of a Subsea Instrument housing generally offers stable temperature and humidity characteristics, easily within the normal operating criteria of industrial instruments and I/O systems. As we have seen, there are also available today, a varied range of equipment for the purchasing engineer to consider without recourse to costly and time consuming special engineering. Using this equipment will enable operators to continue to reduce oil and gas well development costs and substantially ease the use of such systems well into the foreseeable future.

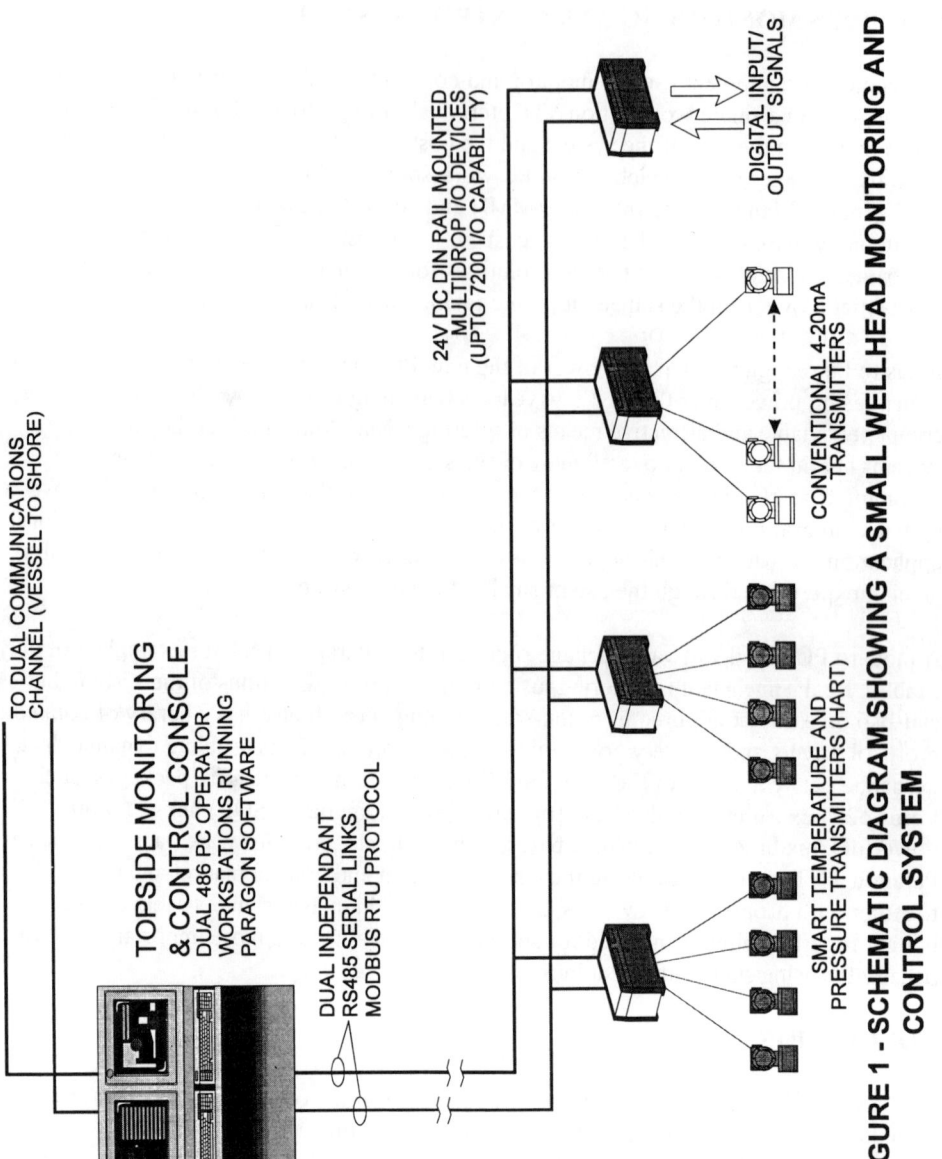

FIGURE 1 - SCHEMATIC DIAGRAM SHOWING A SMALL WELLHEAD MONITORING AND CONTROL SYSTEM

# HYDRAULIC ACCUMULATORS IN THE OFFSHORE INDUSTRY

Eur.Ing. GRAHAM J. MILLER C.Eng. M.I.Mech.E
Technical and Operations Director
Fawcett Christie Hydraulics Ltd
Sandycroft Industrial Estate
Sandycroft
Deeside
Clwyd
CH5 2QP

## ABSTRACT

Hydraulic accumulators have been in existence for some 40 years but the exacting requirements of the offshore industry have driven the technology to new levels.
Quality, reliability in service and minimum weight coupled with higher working pressures have required new ways of thinking and produced specialised product.
This paper looks at the changes that have occured including the current state of the art and examines the technical considerations that face the system designer who has to incorporate the product.

## THE EARLY YEARS

Prior to the development of the offshore industry in the North Sea, there had been very little change in the basic design of hydraulic accumulators.

They had been used in industrial applications, typically machinery such as die casting equipment which had intermittent high flow rate requirements. As a result of this, some development work had been carried out such as fitting large fluid ports but the market was generally satisfied by the range of products then available.

*Volume 32: Subsea Control and Data Acquisition, 211–216.*
© 1994 *Society for Underwater Technology. Printed in the Netherlands.*

In the early 1970's, we became involved in the offshore industry supplying accumulators for use on the B.P. Forties Project for operating ballast valves on the upending system. Supplying the industry in those days was relatively simple with technical specifications limited to one page and any Quality Assurance documentation to a similar size!.

As time progressed, accumulators became used on an increasing range of applications and the market became a significant part of our business.

**OFFSHORE SPECIALISATION**

Towards the end of the 1970's and the development of subsea systems, we supplied what we believe were the first accumulators to be used subset on methanol injection on satellite 'PI' on the North Cormorant field.

These units were adapted for long term installation and were fitted with low permeability bladders to reduce loss of nitrogen gas to a minimum, stainless steel fluid port assembly, metal to metal sealed gas valve assembly with grease filled protective cap and the unit finished with a special paint system.

(You may be aware that PI was retrieved mid 1993 and returned to A.B.B. Vetco Gray, Aberdeen for inspection. With the exception of minor paint problems and the fact that some of the anodes had been painted, all the equipment was in good condition and the bladders exhibited very little marking or deterioration. A paper was presented in December 1993 by Steve Cromer of ABB Vetco Gray at the Subsea '93 International Conference in London).

Further offshore development work was around an accumulator with a one piece shell incorporating a split flange connection and double seals on the gas end to prevent a possible loss of hydraulic fluid (mineral oil) into the environment.  A low permeability bladder was now fitted as a standard requirement. This unit was supplied for several years but discontinued due to cost of interfacing with the system and long manufacturing lead times for the shell, which was available only in one size.

## THE 1980's

Accumulators of various capacities between 10 and 50 litre nominally were now being specified which demanded the use of standard industrial shells fitted with a variety of end connections with double 'O' ring seals now incorporated on the fluid port end.

Technical specifications were becoming more complex, for example:

- System cleanliness requiring the accumulators to be flushed to NAS 1638 Class 6 thereby necessitating the construction of special flushing rigs.

- Special materials of construction with the emphasis on ease of maintenance topside and subsea.

- System pressures in excess of 345 bar which was the limit
  of forged shells and therefore piston accumulators became
  the preferred route here.

## BLADDER ACCUMULATORS versus PISTON ACCUMULATORS

Piston accumulators can effectively be manufactured to any capacity and pressure rating and to an infinite range of bore diameters.

The advantages of each type can be summarised as follows:-

| BLADDER | PISTON |
|---|---|
| – Cost | – pressure/capacity range |
| – tolerance to fluid<br>  contamination | – non catastrophic failure |
| – inertia | – piston position indication |
| – weight | |

The next stage of development was a combination of the above with construction along the lines of a piston accumulator but having internally profiled end caps to accommodate a bladder with its stem in pressure balance. We have manufactured these units for working pressures up to 1500 bar in conjunction with our sister company Oiltech Norway on behalf of Rotator/Saga in a downhole safety valve application.

Whilst this type of unit offers the advantages of the bladder accumulator, it suffers from the increased weight of the piston accumulator.

## FORGED HIGH PRESSURE SEAMLESS SHELLS

Whilst shells for bladder accumulators designed nominally to working pressures of 345 bar are spun, we have now developed hammered shells which are available for M.W.P.'s of 690 and 800 bar (BS7201 Design & Manufacture).

These are available in capacities of 12-50 litre manufactured from chromium molybdenum steel and give a weight saving of around 40% over a similar capacity piston/bladder type.

The manufacturing route can also be used to forge duplex stainless steel to provide seamless shells rated up to 500 bar.

## SIZING TECHNIQUES

With the increase in working pressures, the polytrophic coefficient or 'n' factor needs to be re-assessed as variation from the traditional value of 1.4 will be experienced in practice.

The factor depends on four parameters:

- Maximum & Minimum system pressures
- Flow rate / time
- Temperature
- Draw off volume required
- Stabilisation times between charging/discharging cycles

The calculations can quickly be carried out on behalf of customers using simulation software developed in house by our Group. This gives confidence to 'rule of thumb' techniques which generally oversize the units but we have experienced undersizing by traditional methods in the past.

## QUALITY and RELIABILITY

Accumulators have a good history of reliability in service but retrieval of subsea equipment with suspected quality problems can be expensive in terms of cost and lost production.

Equipment must be fully tested and manufactured to the highest standards before offshore topside and subsea installation.

The heart of the bladder accumulator is the elastomeric bladder which is not simply 'just a rubber moulding' - they are carefully designed to match the interior of shell to ensure long life and no trapping of fluid.

The mechanical and physio chemical properties of each type of rubber is now specified with the focus on the ability of the rubber to resist nitrogen gas permeation. Olaer group uses one source of bladders (part owned Company which does not supply to anyone else) and the same technology provides them for use in aircraft, rocket, naval and other industries demanding total reliability in service.

## ACCUMULATOR APPLICATIONS

Amongst many applications, accumulators are successfully used in:
- Emergency shut down valves
- Soft landing systems
- Subsea Control Pods
- Acoustically operated systems

They are also now being used as a replacement of the spring for spring return actuators. Benefits gained include compactness, reduced weight and simplified maintenance.

In conclusion, accumulators have demonstrated their versatility, both topside and subsea, in the offshore industry.

# SUBSEA DISTRIBUTED DATA ACQUISITION SYSTEM

J RAMSHAW
Research and Development Group
FSSL Ltd
Monarch House
Victoria Road
London W3 6UL
United Kingdom

## SUMMARY

This paper describes a Subsea Distributed Data Acquisition Bus for the Subsea Hydrocarbon Production Control environment. This will have many advantages in terms of flexibility, maintainability, fault tolerance, data confidence and overall cost.

An outline is given of the development project describing the contents of each of its four phases. The product of this project will be a standard definition of the bus which will be non-exclusive, non-proprietary and vendor independent.

A system overview is presented with descriptions of typical components. An outline of the relevant areas of the MIL STD 1553B bus standard is given. This bus has been selected to be the basis for the subsea design. Other topics covered in the overview include typical system configurations, connection methods and power and signal distribution. Options are proposed for further investigation during the development project.

The final section of the paper gives a brief description of three of the many possible applications of the proposed subsea digital communication and power distribution bus.

## INTRODUCTION

Subsea control systems for the production of hydrocarbons are typically made up of an operator interface on a platform, a subsea communications and power distribution system, a number of subsea control modules and subsea instrumentation.

Hydraulic power is supplied to the control module. On command from the surface solenoid piloted electro hydraulic valves are operated via the communications network that direct this supply to control actuators on the well Xmas tree.

Volume 32: Subsea Control and Data Acquisition, 217–236.

Instrumentation mounted on the tree, and also sometimes downhole, provide signals that are digitised in the control module. This data is passed to the operators console on the surface to provide information on the state and performance of the well. Figure 1 is a block diagram showing the typical existing data acquisition system set-up. Each sensor is connected by a dedicated jumper and connector back to the control module. Each analog channel therefore requires a connector mounted on the control module which adds to its overall weight and size.

An alternative proposed system of data acquisition for subsea control systems is illustrated in Figure 2. This involves the use of a digital communication and power distribution bus, and digital sensors. The subsea control module is connected to the sensors via the dual redundant bus. Systems of this sort are finding increasing use and acceptance in industrial process and control applications. These are commonly referred to as Fieldbuses.

The major difference between the existing and proposed subsea data acquisition systems is that in the new system the analog signals are converted to digital data in the sensor as opposed to being transmitted back to the control module for conversion. The digital sampled data is then requested from the sensor by the control module. Transmission of digital data is far more tolerant of noise and line degradation than for example 4 to 20 mA, so the repeatability of the proposed bus system is much improved. Confidence in the data received at the module and hence at the surface is therefore increased.

By using digital sensors linearity, zero and offset can be corrected for within the local digitisation circuitry if required. This removes the task of keeping a particular sensors calibration data associated with the analog channel it is connected to and ensuring that the correct data is downloaded to the respective control modules. In a large field this can be fairly complicated. The fact that corrected data is generated by the sensors simplifies the algorithms within the subsea control module and also the surface computer configuration.

Another of the main advantages of the Subsea Distributed Data Acquisition Bus (SDDAB) concept is its flexibility. A vast array of different types of sensor or subsystem can all be used with a standard interface to the control module. This allows a far more standard approach to subsea control systems than has previously been possible. The replacement of the many analog connectors in existing designs with a few digital bus connectors can significantly reduce the size and weight of control modules. This becomes more and more important with the trend for increasing water depth and ROV installation.

Figure 1. EXISTING DATA ACQUISITION SYSTEM

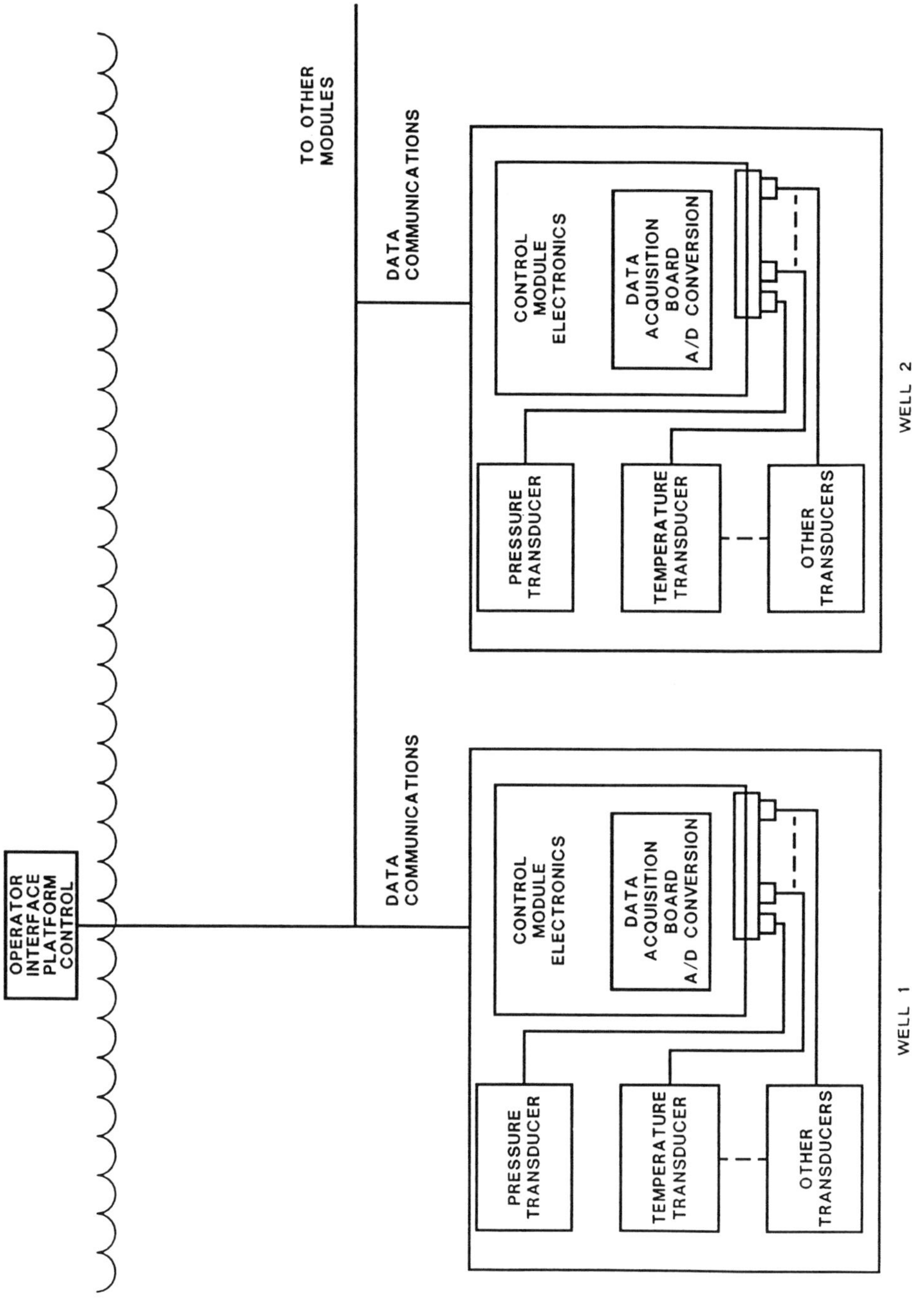

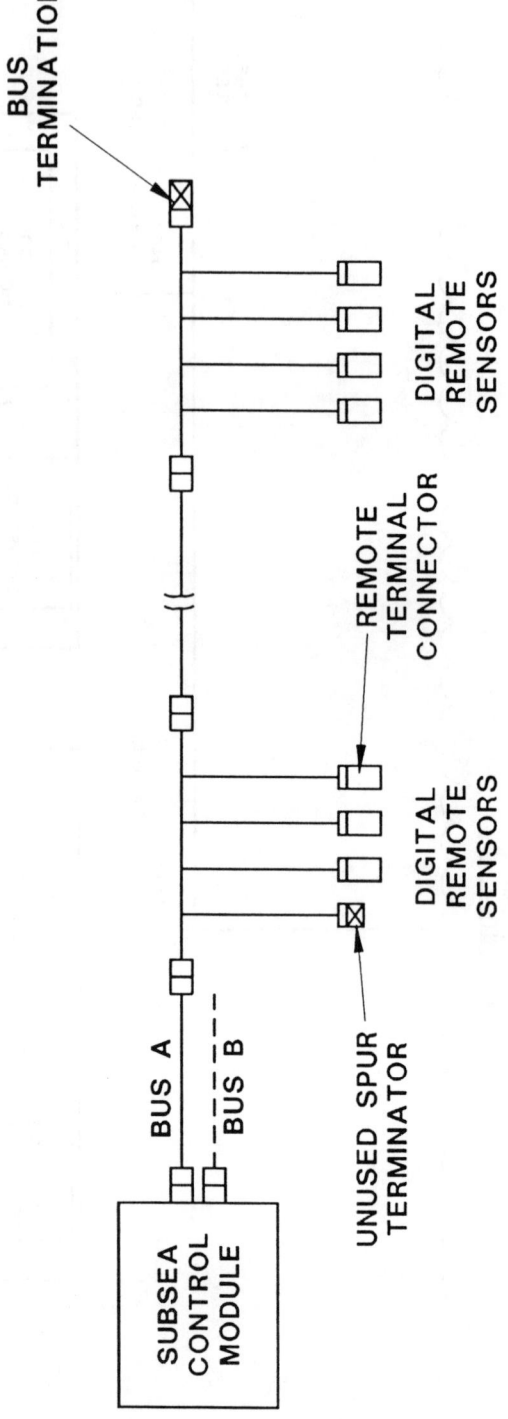

Figure 2. A BUS BASED DIGITAL DATA ACQUISITION SYSTEM

# SUBSEA DISTRIBUTED SENSOR DEVELOPMENT PROJECT

The aim of the Subsea Distributed Sensor Development Project is to develop a digital communication and power distribution bus suitable for the Subsea Hydrocarbon Production Control environment. The project is divided into four phases.

## Phase 1 - Conceptual Study

The first phase of the development is a study to evaluate available technologies, the feasibility of the concept, and to propose possible system designs.

## Evaluation of Existing Standards

An evaluation of the suitability of existing and proposed digital Fieldbus standards has been undertaken. The buses included in this study were:

- FIP UTE Standard C46-600 series (ref 1)

- PROFIBUS Standard DIN 19245 Pts 1 & 2 (refs 2 & 3)

- ISA FIELDBUS Standard (part issued/part draft) (ref 4)

- MIL STD 1553B (ref 5)

The aim of this evaluation was to identify either a single existing standard which may be implemented in full for subsea systems or a reduced standard, subsets of which may be implemented for a subsea system. Each standard was assessed under the headings of the OSI model for open systems interconnection, with particular reference to the physical and data link layers.

The three industrial fieldbus systems, FIP, PROFIBUS and ISA, have been designed for process control applications of a complexity far beyond any envisaged subsea requirement. They are based upon complex systems for data/message handling with intelligent interactive remote terminals and the facility for multiple active bus masters. This along with the difficulty of high reliability implementation, associated with all subsea electronics, severely limits their adaptability to the subsea requirement.

In defining a basic digital data link with simple message formats the MIL STD 1553B standard best lends itself to adaptation for the subsea application. The proposed subsea data acquisition system is therefore based upon the MIL STD 1553B standard with modifications to match the environment and provide the level of functionality required. A brief outline of the main features of MIL STD 1553B is included below.

High reliability implementation of a bus system based on the MIL STD 1553B standard is simplified by the availability of fully qualified components and application specific building blocks.

**Conceptual Design**

The conceptual design task within Phase 1 investigates the feasibility of implementing a SDDAB and proposes possible system concepts and configurations. An outline of some of the proposals is included in the system overview.

**Phase 2 - Detailed Design**

The preferred conceptual design will be selected and considered in detail in Phase 2 of the development project. The aim of this phase is to develop electronic and mechanical hardware and it includes the detailed design, prototype and testing of typical system components. Software is written to allow the components and system to be fully functionally tested.

**Phase 3 - System Integration and Testing.**

On completion of the detailed design the components are integrated into a working system. A qualification program designed to test the hardware and software over a wide range of operating conditions is carried out. The performance of the system will be evaluated and any required changes made to the design. The system performance data collected will confirm the maximum and typical capability of the bus and will be used as the basis for the design of future systems.

**Phase 4 - Standard Release.**

The concluding phase of the project will release a document defining the Subsea Distributed Data Acquisition Bus developed within this project as a proposed standard for use in the Subsea Controls for Hydrocarbon Production Application. It is envisaged that this should be a non-exclusive, non-proprietry and vendor independent standard.

## COMPONENT OVERVIEW

**Typical System Components**

The Subsea Distributed Data Acquisition Bus (SDDAB) can be considered in terms of three major components, the Master Terminal, the Remote Terminal and the Interconnection System.

**The Master Terminal**

All bus based communication systems require at least one entity on the bus to control the communications traffic. This entity can either be virtual or physical. The SDDAB makes use of a physical entity, the Master Terminal, to control the entire bus in a master slave multi-drop configuration.

The Master Terminal is positioned within the control module and is the gateway for the Subsea Control System to the SDDAB. The major functional requirements of the Master Terminal can be summarised as:

- Initiation of all communication on the bus.

- Collection of data from the SDDAB.

- Hand-off of SDDAB data to the rest of the Subsea Control System.

- Conditioning and distribution of power to the SDDAB.

- Condition monitoring of the dual redundant power and data paths.

- Automatic or surface commanded switch over between the dual redundant power and signal paths.

### The Remote Terminal

The Slave entities on the SDDAB are termed Remote Terminals. These would typically be the sensors on the bus. Each Remote Terminal has a unique address on the bus to allow the Master Terminal to communicate with it.

Figure 3 illustrates the block diagram of a typical Remote Terminal. The device can be considered as being made up of three functional blocks.

### The Bus Interface

The physical and logical connection to the bus is called the Bus Interface. Communications and Power are extracted from the active bus. Messages with the correct address are decoded and checks are made to ensure the integrity of the received data. This data is then handed off to the control logic.

Responses to the Master are generated by the control logic and are passed to the bus interface. These are converted into the correct message format and are sent on the bus.

### The Control Logic

The functionality of the Remote terminal is produced by the control logic. This is the intelligence within the terminal.

Messages are received from and sent to the Master Terminal via the bus interface. The received messages are interpreted by the control logic and are acted upon. An example message would be a request for some data. The control logic assembles the requested data and passes it on to the bus interface for transmission on the bus.

The control logic is responsible for collecting and temporary storage of data from transducer interfaces. Multiple data points will be available, depending upon the particular functionality of the Remote Terminal in question and the control logic complexity. Data correction can be performed on the raw data by the logic before it is transmitted across the bus.

**The Transducer Interface**

The transducer interface consists of the circuitry required to condition and convert the signal from the transducer into digital data. The control logic will normally control the conversion process and store the resulting data samples.

The type of transducer will dictate the complexity of this circuitry. A bridge based pressure transducer will have a very simple interface when compared with a multiphase flowmeter for example.

The transducers will generally be included in this functional block.

**Interconnection System**

The interconnection system covers all aspects of how the bus master terminal is connected through a bus to the slave Remote Terminals.

Some of the main functional requirements of the interconnection system are summarised below.

- All components used in the interconnection system must be qualified for the Subsea environment.

- The system must allow for subsea mating and demating of connectors preferably without interruption to the bus.

- Both diver assisted and ROV operation should be supported for installation, maintenance and modification of the bus.

- The system should make full use of the dual redundant features of the selected bus.

- It is important that with a flexible bus system Terminal addition or reconfiguration of the system has minimum effect upon performance.


**THE MIL STD 1553B BUS**

Military standard 1553B was developed in the mid-1970's as a time division command/response multiplex data bus for military aircraft.

Since then it has become the accepted standard for avionics applications and in the last 10 years it has found applications in naval vessels, ground vehicles and in some non military areas. It has become the prominent standard for military medium speed communications and hence is fully supported by a wide range of electronic hardware manufacturers. Because of the military/safety critical application, all hardware is available with full high reliability MIL STD 883B qualification. Both master and remote terminals can be implemented using single package hybrids.

MIL STD 1553B differs from the other fieldbus standards in that it only defines what in the other systems are termed the physical and lower data link layers. Higher data link functions (i.e. complex error checking) and the application functions (i.e. bus arbitration, cyclic polling of slaves etc) may be implemented in whichever way best suits the application using custom software written for the host CPU. The complexity of the system is therefore fully defined by the system designer to satisfy the ultimate design objectives.

**Physical & Data Link Layers**

The MIL STD 1553B physical and data link layer equivalent may be summarised as follows:

- Transmission medium is shielded twisted pair. Length dependant on cable attenuation characteristic.

- The linear bus is terminated at both ends with characteristic impedance.

- Maximum transmission speed is 1Mbit/s.

- Coding is Manchester II biphase, half duplex, asynchronous transmission.

- Each transmitted word consists of sixteen bits, preceded by a 3 bit synch. pulse and terminated with a single bit parity check.

- Transmitted information consists of three types of word:

  - Command word - containing a remote terminal address and mode code

  - Data word - up to 32 16-bit words always following a command word or status word

  - Status word - containing remote terminal address and eleven status bits

- Messages may be one of six formats:

  - Bus controller to remote terminal transfer

  - Remote terminal to bus controller transfer

  - Remote terminal to remote terminal transfer (under command of bus controller)

  - Mode command without data word (status word request)

  - Mode command with data word

  - Broadcast commands

- Sole control of information on the bus resides with the bus controller (the master terminal).

- Bus control may be passed between master terminals but only one can be active at any one time.

- Redundancy is supported using twin serial buses. All integrated hardware contains twin transceivers.

**Implementation/Hardware**

Implementation of MIL STD 1553B requires system specific hardware for the physical and lower data link layers. The master terminal operates under control of a host microprocessor. This could be achieved either using the Subsea Control Modules processor, for less speed critical applications, or by a installing a microprocessor dedicated to the task. The remote terminals control logic will be implemented using either a microprocessor or programmable logic depending on the particular terminals requirements.

Single hybrid implementations of complete MIL STD 1553B terminals are available from a number of manufacturers with multi-sourced options.

Reduced complexity hybrids implementing only a limited range of MIL STD 1553B modes are also available.

All hybrids are available to MIL STD 883B screening and are suitable for subsea installation.

**MIL STD 1553B Usable Features for the SDDAB**

The following MIL STD 1553B features would be advantageous in the context of a typical subsea system:

- Manchester encoding with it's associated noise immunity and it's potential for combined power and communications transmission. (NB. Combined power and communications is not standard within 1553B).

- Transmission speed up to 1 Mbit/s.

- Simple frame structure with a short synch pulse and parity bit transmitted with every sixteen bit word.

- Low data overhead allows high transmission speed for short data transfers.

- Acknowledged data transfers with hardware check for parity, transmission errors and word count errors.

- Data bus redundancy with twin transceivers in all terminals. Twin master stations may be configured on the bus with one master looking like a remote terminal until bus control is passed to it.

- Absence of predefined application layer allows simple system specific application programs to be written to suit for example a limited master-slave system (as typically used in subsea applications).

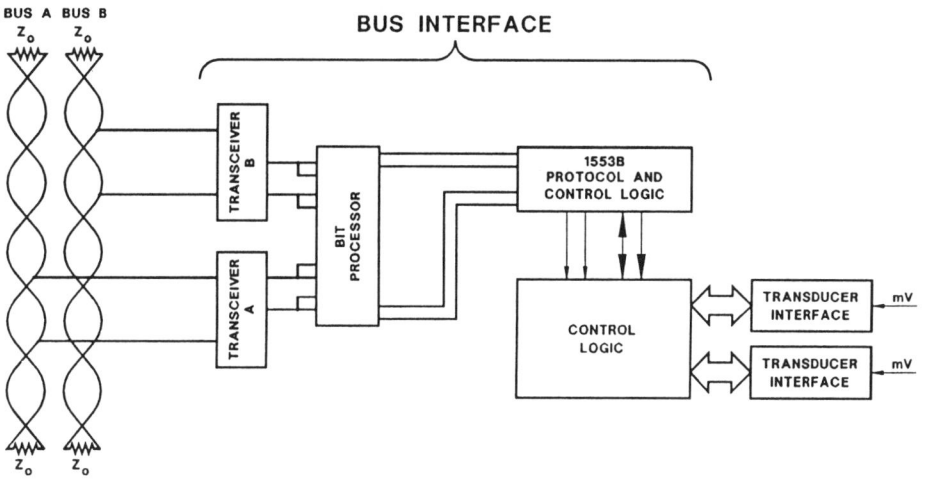

Figure 3. A Typical Remote Terminal Block Diagram

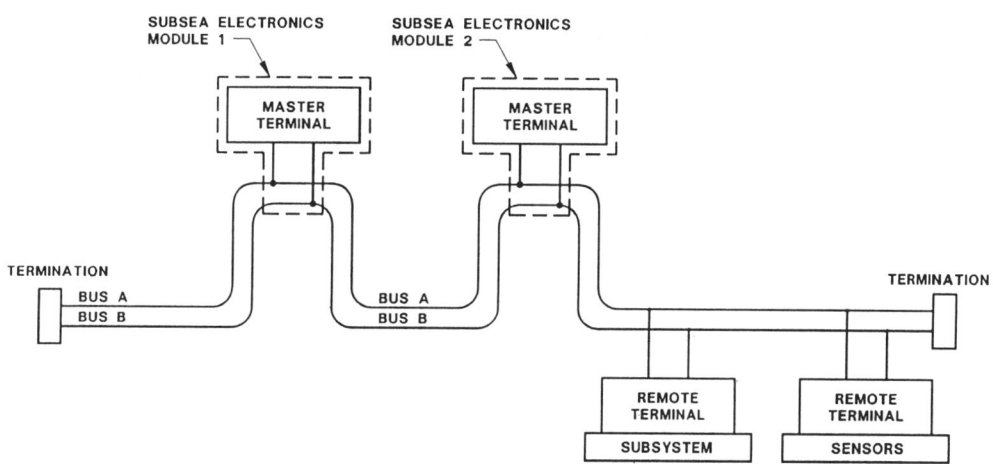

Figure 4. A Redundant Master Terminal System

- Both master and remote terminals may be compactly implemented in single hybrid packages.

- All system hardware available from multiple manufacturers and qualified to MIL STD 883B.

- System fully proven and widely used in highly critical applications (ie. military avionics).

## SYSTEM OVERVIEW

### Typical System Configurations

By its nature the Subsea Distributed Data Acquisition Bus is very flexible and many different topologies can be envisaged for different applications. For subsea well and template control three main categories can be identified dependent upon the level of redundancy required, the single master sytem, dual redundant master system and multiple bus system.

### Single Master System

The basic system is based around a single bus made up of one master terminal and several remote terminals. The master terminal would typically be housed within the Subsea Electronics Module and be integrated with the local control and data acquisition system. This set up is illustrated in Figure 2. There is one level of redundancy within this system. The Master is connected to the Remote Terminals via the dual redundant bus. Only one path of the bus is used at any one time under the control of the Master Terminal. Switching between paths can either be achieved automatically or by command from the operator at the surface. This would typically occur if communications was lost or corrupted and if the load on the power supply went outside predefined limits.

### Dual Redundant Master System

Redundancy in the system can be increased by utilising dual redundant Master Terminals as indicated in Figure 4. In a fully dual redundant electronics system the two Master Terminals would be housed one in each Subsea Electronics Module within the control module. An alternative application would be to house the master terminals in different control modules allow both access to the same bus. In a dual redundant master system only one of the terminals acts as bus controller at any one time. The second master can either act as a Remote Terminal or can be a bus monitor. By monitoring the bus the redundant master can receive the same information as the bus controller allowing the second module to keep its data up to date whilst in standby. This facilitates more efficient switch over between the two units. The selection of the active bus controller is carried out by command from the surface in association with the active SEM or control module. If both masters are fitted within the same SEM the switch over could alternatively be handled by the local intelligence.

## Multiple Bus Systems

Both single and dual redundant masters can be implemented for multiple bus systems. As each bus can only have 32 entities it is envisaged that for some systems multiple busses will be required from the control module.

Figure 5 shows a block diagram of a system using two busses. Bus #1 is connected to only one critical sensor to ensure the highest reliability and data collection speed whilst the other less critical terminals are connected to bus #2. The terminations of each bus are designed so that their removal allows the addition of sensors and system reconfiguration.

Multiple busses add a great deal of flexibility to the Subsea Distributed Data Acquisition System. The potential for very high functionality can be built into a control module at a very low relative cost. This facilitates the design of a field independent control module and multi-well control.

## Connection Methods

There are three types of connector that could be used for the SDDAB. The second phase of the development will investigate more fully the suitability of each type of connector.

The SDDAB could be implemented using subsea mateable pin to pin connectors, inductive couplers or capacitive connectors. All of these techniques are compatible with AC power distribution. The pin to pin connectors are also compatible with DC distribution.

Pin to pin connectors offer the least technical risk to the development but the other techniques support greater bus security and functionality. As both inductive and capacitive couplers can be mated and unmated subsea with the power on they offer the facility for reconfiguration of the system without having to shut down. This becomes increasingly preferable as the functionality of the control module increases. For example if a module is controlling a template of multiple wells it would not be acceptable to loose all data acquisition so that sensor can be replaced on the bus.

The other advantage of inductive and capacitive coupling is that they provide for a higher level of fault tolerance than the conductive type. This is because by their nature they isolate the terminals from the bus.

## Power and Signal Distribution

Power and signal can be distributed in one of two ways on the SDDAB, either separately using two pair cable or combined in a single pair of conductors. The advantage with the combined power and signal set-up is that both component size and cost are reduced.

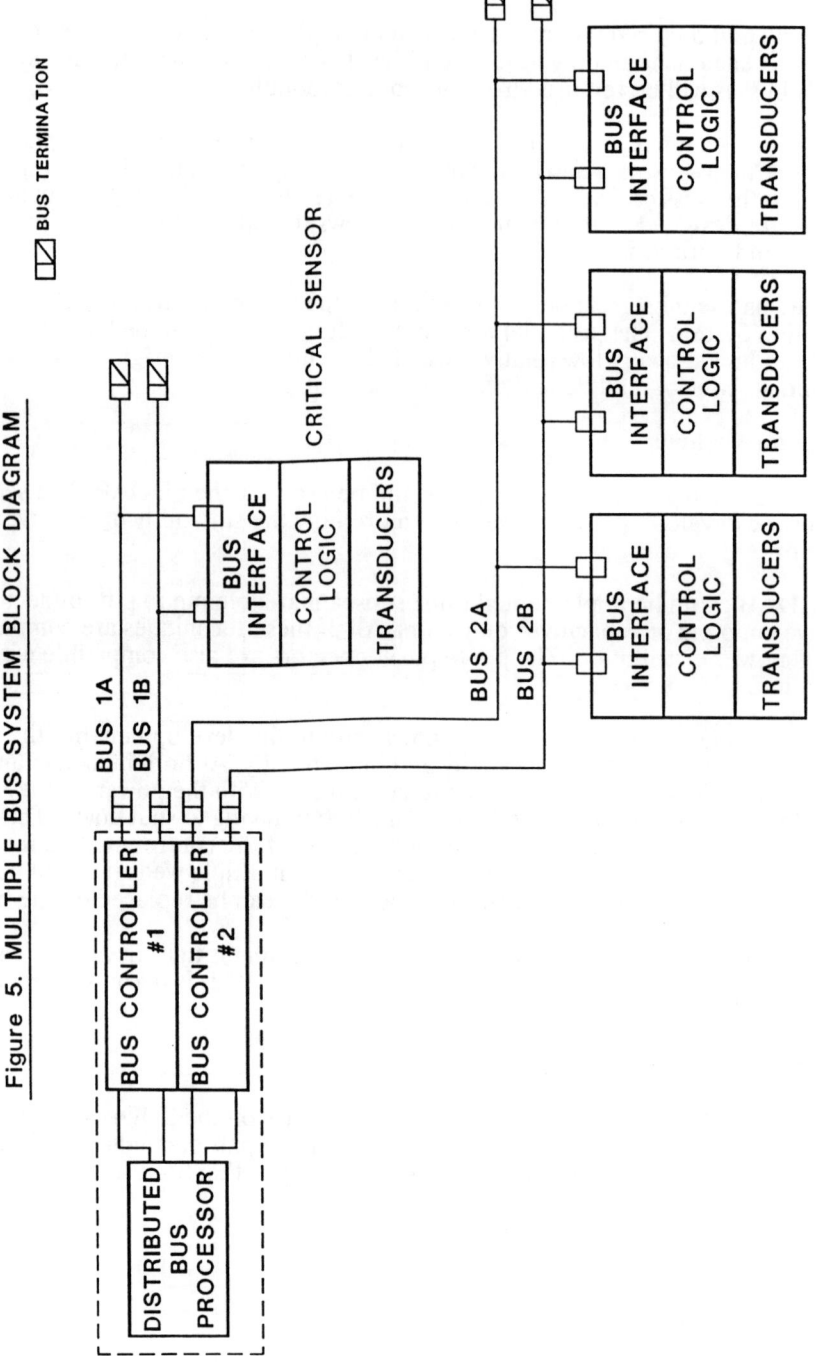

Figure 5. MULTIPLE BUS SYSTEM BLOCK DIAGRAM

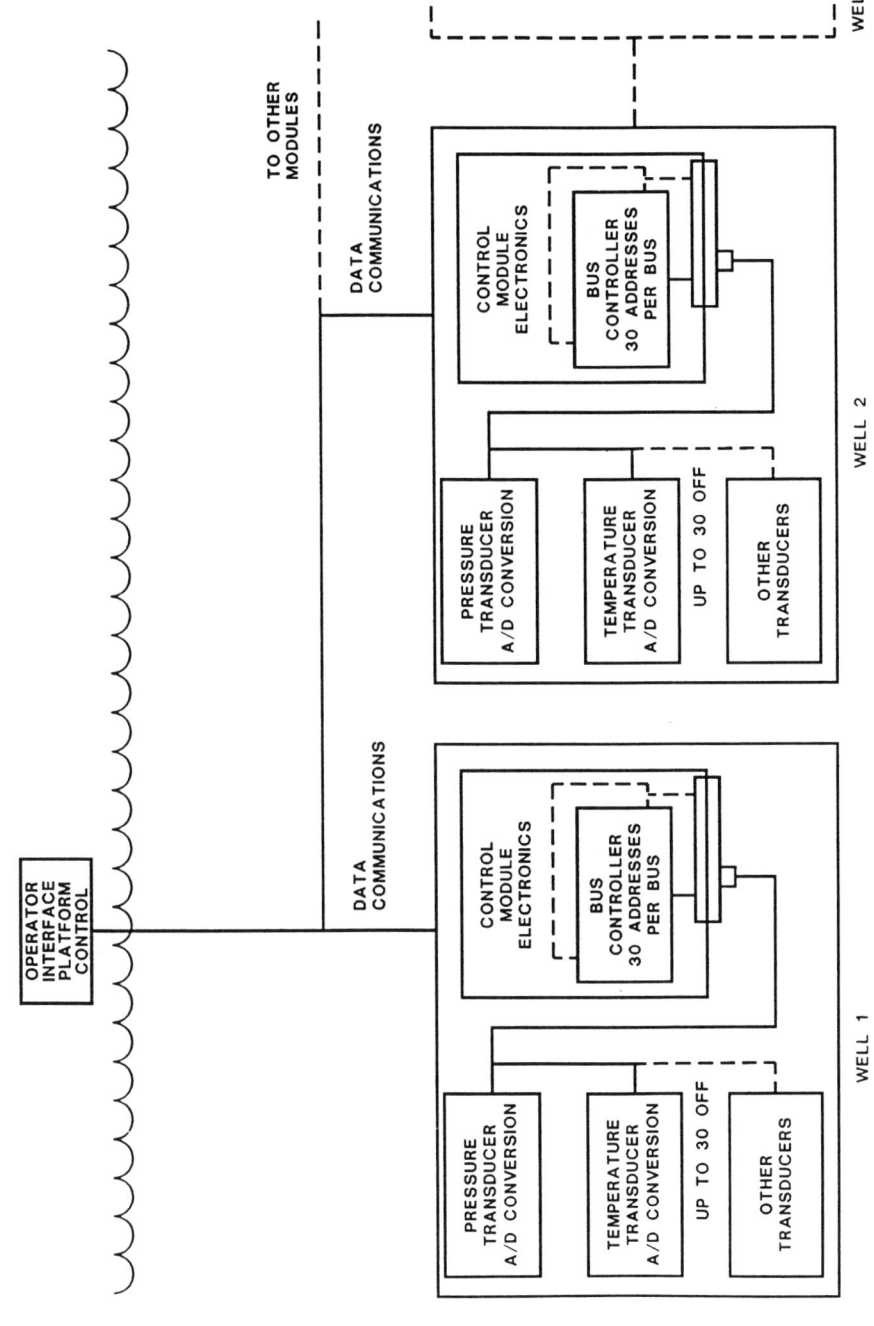

Figure 6. DISTRIBUTED DATA ACQUISITION SYSTEM

Figure 7. FULLY DISTRIBUTED DATA ACQUISITION SYSTEM

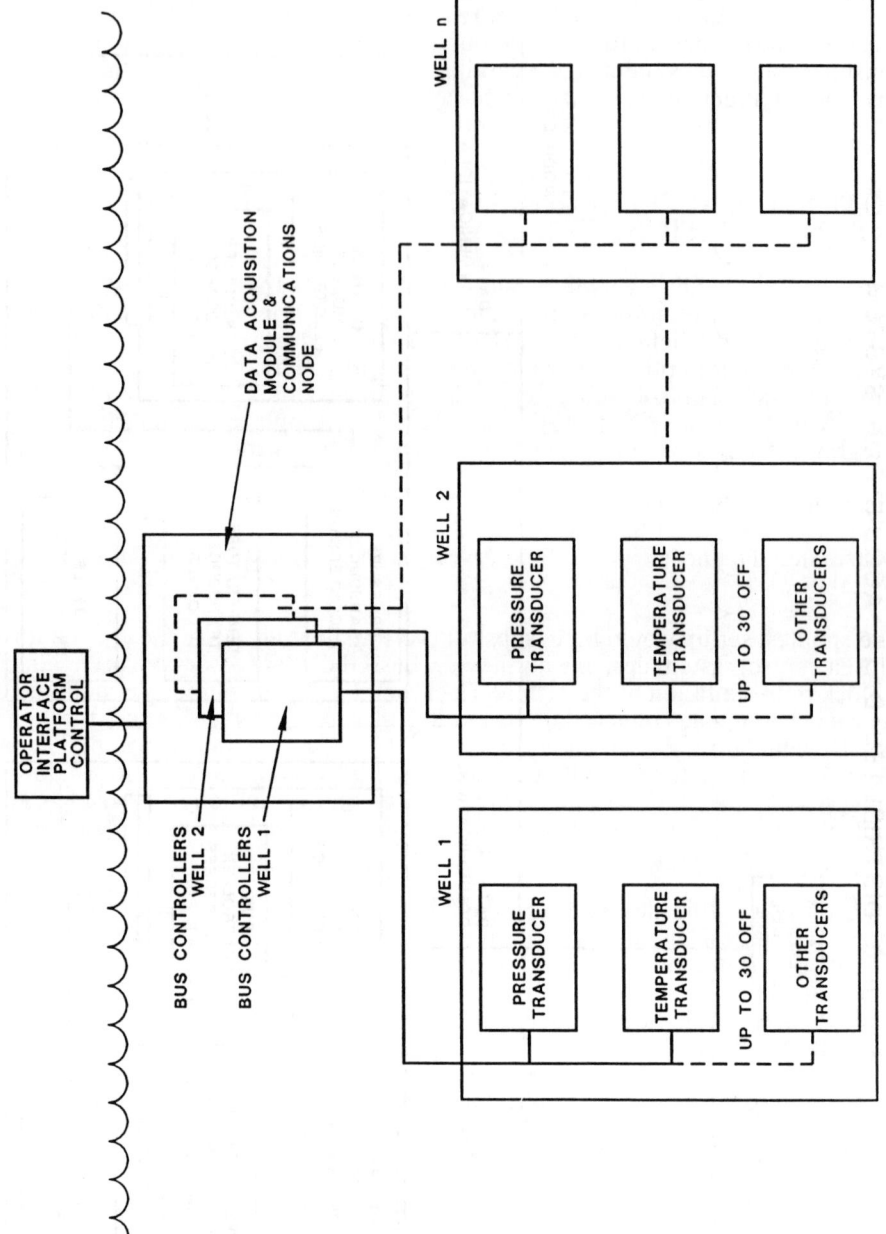

The proposed system for development is combined power and signal using AC power. The use of AC will enable more efficient power transfer allowing more sensors to be connected to the same bus and a reduction in the overall power consumption of the system. The overall power consumption is important because it has a direct affect upon the sizing, and hence cost, of the power distribution components and network for the overall Subsea Control System.

# POTENTIAL APPLICATIONS OF A SUBSEA DIGITAL COMMUNICATIONS BUS

There are potentially a very large number of different applications of the subsea digital communications bus outlined in this paper. A brief description of three applications in the field of Subsea Hydrocarbon Production Control are presented below. Examples of other fields that the technology could be applied to include Subsea Process Control, Subsea Fiscal Metering and all other aspects of Subsea Data Acquisition and Control. The distributed philosophy also lends itself to integration of these different functions into a single control system.

## Distributed Data Acquisition

Distributed data acquisition subsea is seen as the most likely first application of the digital bus.

The simplest set-up would be to substitute the analog sensors in the conventional subsea control system design with the new bussed digital sensors. Figure 6 illustrates a block representation of the system. Each well has a control module mounted on or near it. The module collects data from the Xmas tree mounted sensors through the bus. Downhole gauges can either be incorporated on to the bus downhole or, more likely, have a remote terminal at the well head interfacing to the standard downhole gauges. The control of the well is carried out in the conventional way using solenoid control valves within the control module. This allows the benefits of digital sensors whilst still retaining the established subsea control philosophies.

An extension to this idea would be the Fully Distributed Data Acquisition System shown in Figure 7. In this application there is a central subsea data acquisition module that is monitoring several wells. More than one bus can be assigned to a well depending upon the actual field requirements. An example of where this system could be used would be a field or template of multiple direct or piloted hydraulic controlled wells.

Incorporating the electro hydraulic control for the multiple wells into the central data acquisition module, and perhaps duplicating this to provide dual redundancy, leads to a potentially very cost effective control system.

The philosophy of multiple well control using dual redundant control modules is not a new one. Several developments in the Norwegian sector of the North Sea use this principle. The utilisation of the digital bus increases the flexibility, accuracy, repeatability and hence cost effectiveness of these systems.

Figure 8. FULLY DISTRIBUTED DATA ACQUISITION & CONTROL SYSTEM

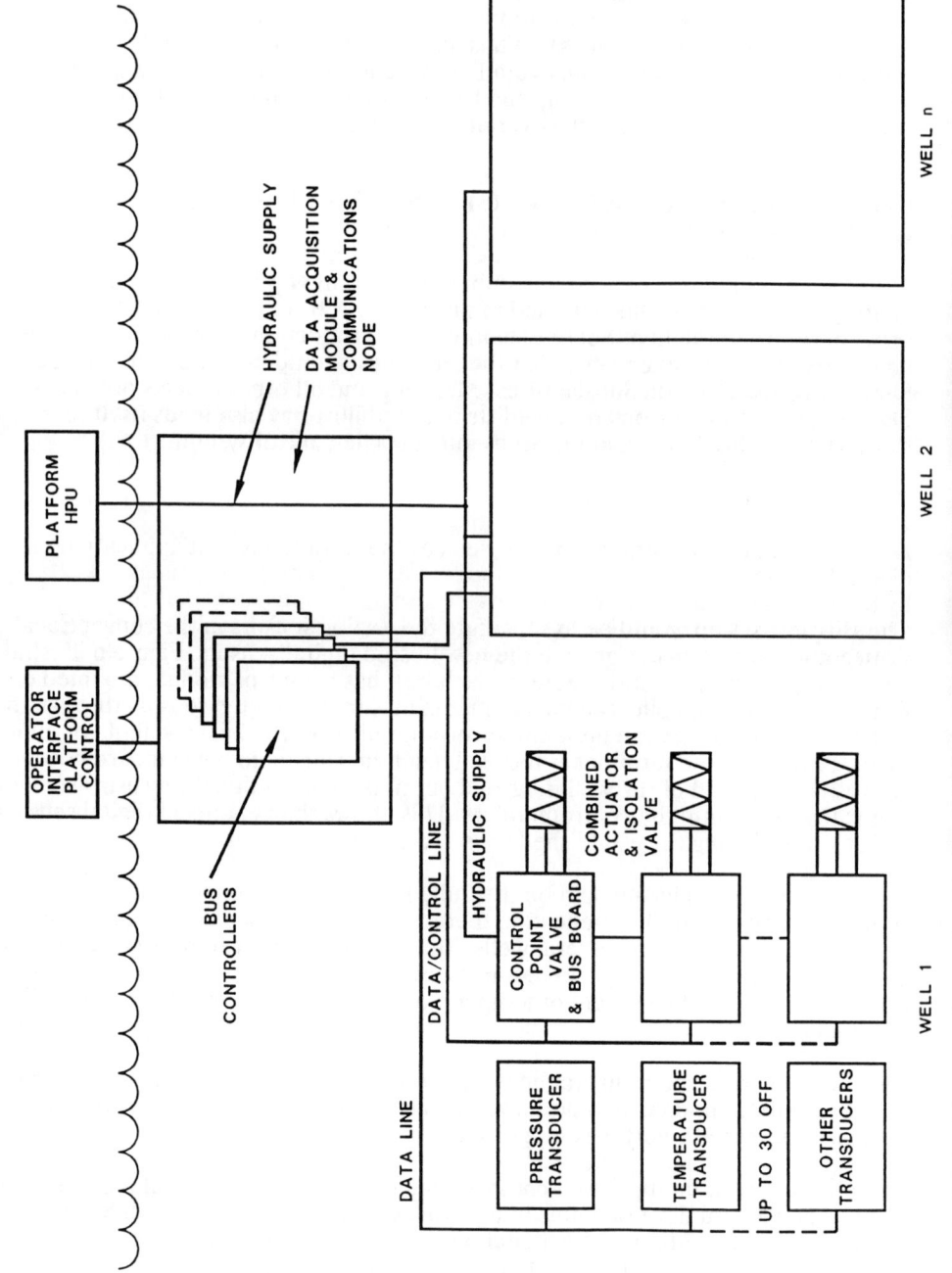

## The Standard Subsea Control Module Concept

A major application of the flexibility made possible by using the SDDAB is the specification of a "standard subsea control module" for a particular operator.

A significant proportion of the cost of a Subsea Control System is non recurring engineering. Designing a control module with sufficient functionality (in terms of electro hydraulic control valves) and a suitable number of SDDAB controllers to cope with a standard well has many advantages. As well as reducing the non recurring engineering costs for a particular development associated equipment and tools for deploying and maintaining the system can be field independent.

One of the factors that has hindered any supplier offering a truly operator standard product has been the fact that instrumentation requirements change over time. New field developments have to be able to move with the up and coming subsea technology to enable the most cost effective production of hydrocarbons. As the SDDAB makes no assumption about the functionality of the instrument or subsystem attached to it any future devices should be easily interfaced to the bus based standard control module.

## Distributed Control and Data Acquisition

The SDDAB is a digital communication bus. The communication facilities offered by the system could easily be used for control applications as well as the data acquisition tasks already discussed. Figure 8 shows a block diagram of the fully distributed data acquisition and control philosophy.

The surface to subsea communications link is to a central data acquisition and control module. This contains the majority of the subsea control intelligence and several subsea bus controllers. This module could control only one well but more typically could control several. Hydraulic supply is passed through this module with associated filters, isolation valves and possibly electrically held fail-safe valves.

The busses now connect the control module not only to digital sensors but also to small control packages mounted on the actuators themselves. Hydraulic power supply is passed around the tree in a similar fashion to the SDDAB. The control package remote terminal would typically comprise of a small electronics package containing the bus interface, control logic and solenoid driver circuit, a hydraulic pressure transducer and a solenoid piloted control valve. The whole package would be small and light enough to facilitate easy installation and retrieval using ROVs.

This Distributed Subsea Control System would be the ultimate in the compromise between standardisation and flexibility. Very different control solutions on different fields can be achieved using standard building blocks. Modifications and upgrades to systems can be carried out by using the flexibility inherent in the bus design. As long as the basic interface specifications are met any new devices can be easily backward compatible and all of the building blocks and associated tools can be field independent.

## ACKNOWLEDGEMENT

FSSL wishes to thank Shell UK for their assistance and advice in forwarding this work.

## REFERENCES

1.    UTE-C-46-601 to 607, FIP Fieldbus Standards for Factory Instrumentation
      Protocol

2.    DIN 19245 Part 1, Process Fieldbus Data Transmission Technique Medium
      Access Methods & Transmission Protocols, Service Interface to the
      Application Layer, Management

3.    DIN 19245 Part 2, Communication Model, Services for User Application,
      Protocol Specification, Coding, Data Link Layer Interface, Management

4.    ISA-S50.02-1992, Fieldbus Standard for use in Industrial Control
      Systems.Part 2: Physical Layer Specification & Service Definition

5.    MIL STD 1553B Notice 1/2, Digital Time Division Command/Response
      Multiplex Data Bus

# SUBSEA MULTIPHASE FLOW METERS

E.H. LUNDE and J.G. HOSETH
Development Department
Kongsberg Offshore a.s.
P.O.Box 1012
N-3601 Kongsberg
Norway

## ABSTRACT

Until recently multiphase flow meters have been at an infant stage of development. Only a handfull of meters have been field tested, and then only in a limited amount of onshore / offshore locations and applications . However, papers on the subject of subsea multiphase metering has previously been presented at various venues, this has been described elsewhere, Ref. /2/, but no subsea multiphase flow meter has yet been field tested.

However, the successfull application of the KOS Multiphase Flow meter (MCF) in onshore installations have led to an increased interest in applying these meters also for subsea developments. The potential for cost reductions resulting from reduced capital expenditure as well as reduced operational cost are substantial.

The added complexity resulting from marinization of a technology which has yet to be fully field proven and accepted, poses great challanges. Challanges such as;  space limitations, installation method (permanent or insert), location (flow prediction), reliability and availability,  power and signal transfer, maintenance, intervention, and last but not least - operational requirements and expectations.

This is illustrated with relevant examples of design solutions for the MCF, which is now being prepared for subsea applications.

## INTRODUCTION

Kongsberg Offshore a.s (KOS) in cooperation with Shell Research and A/S Norske Shell, has completed the first two stages of development for a surface multiphase flowmeter, named the Multi Component Flowmeter (MCF). A number of these meters have been tested in the laboratory, as well as in onshore field installations, with very good results. The test results has been described elsewhere, Ref. /4/.

Some of these meters are now permanently installed in the field, where they are being used for continous well monitoring and surveillance, as well as for reservoir management.

Volume 32: Subsea Control and Data Acquisition, 237–252.
© 1994 Society for Underwater Technology. Printed in the Netherlands.

**Multiphase flow conditions**
During multiphase flow the distribution of liquids and gas across and along the pipe are not homogeneous, but varies with time even when the flowrates for liquid and gas are fixed. The variation in distribution increase considerably with changing flowrates, and various distinct flow regimes occures. Flow regimes are also dependant on the piping installation, i.e. vertical or horizontal, flowrates, obstructions etc.
Multiphase pipe flow is thus classified in different flow regimes, as described in the "Mandane Flow Map" for multiphase flow in horizontal pipes, **Refer Figure 1.**
The type of flow regime occuring in a pipe is also dependent on operating conditions, fluid properties, pipe inclination and upstream/downstream conditions.
Flow conditions at the various locations can be dramatically different. Some wells are "clean" oil producers with little water, sand, wax etc. Others are more complex with surging gas production and / or surging watercuts. Others again have high sand production, and some have problems with waxy crude.
In addition the flow patterns in the multiphase flow lines can vary. One minute being a well mixed fluid, the next a slugging flow. Some wells being natural producers, others being dependant on gas lift or engineered pressure maintenance (gas / water injection) for flowing. However, the most common flow regime in multiphase flowlines is intermittent flow, or slug flow as it is also known. This is a flow condition generated by the natural gravity separation process in multiphase flow. Slug flow is a naturally "stable" flow condition. This is also the flow regime which the flow meters must be capable of handling. The required operating envelope for multiphase flowmeters has been described elsewhere, Ref. /3/.

**Measurement technology**
The operating principle for the MCF meter has been described  elsewhere, Ref. /1/, but a brief introduction is given below.
When utilizing the MCF for measurement of the multiphase flow, the liquid and gas flowrates are metered through detecting fluid level and velocity using an intrusive sensor installed in the flowline. Pressure and temperature measurements are also provided by local pressure and temperature transducers.
The MCF consists of a pipe unit complete with an intrusive sensor arrangement comprising several rows of capacitor plates. The spool piece is installed in a horizontal pipe. The sensor arrangement is installed vertically into the pipe unit in line with the flow. Each sensor measures the apparent dielectric constant for the liquid (oil and water) and the gas. **Refer Figure 2.**
The actual flowrates are being derived from liquid hold-up measurements which combined with velocity information obtained from cross correlation circuitries, are fed into a PC based processing unit containing the necessary equations and algorithms for determination of flowrates.
The meter currently operates in slugging flow at gas volume fractions between 30 and 98 % at watercuts upto 40 - 60% (oil continuos emulsions). No mixing or homogenizing of the flow is required.

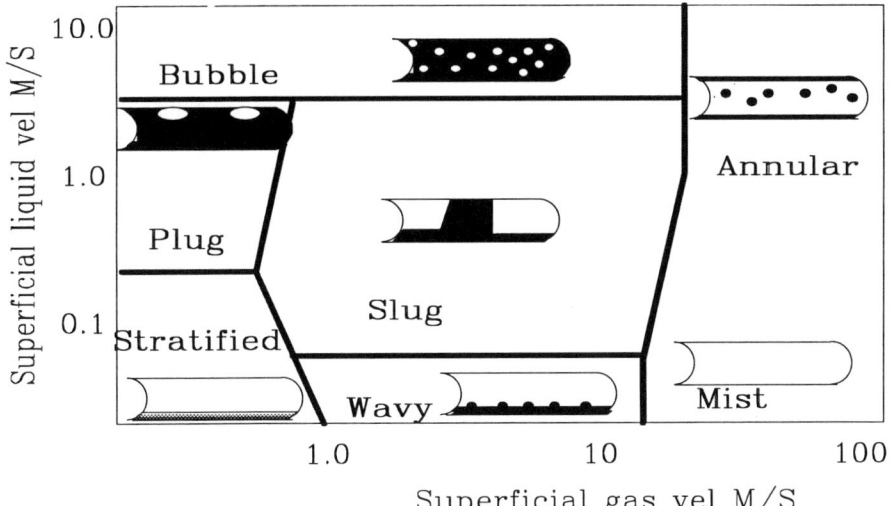

FIGURE 1.

**MANDHANE FLOW PATTERN MAP
FOR MULTIPHASE FLOW IN HORIZONTAL PIPES.**

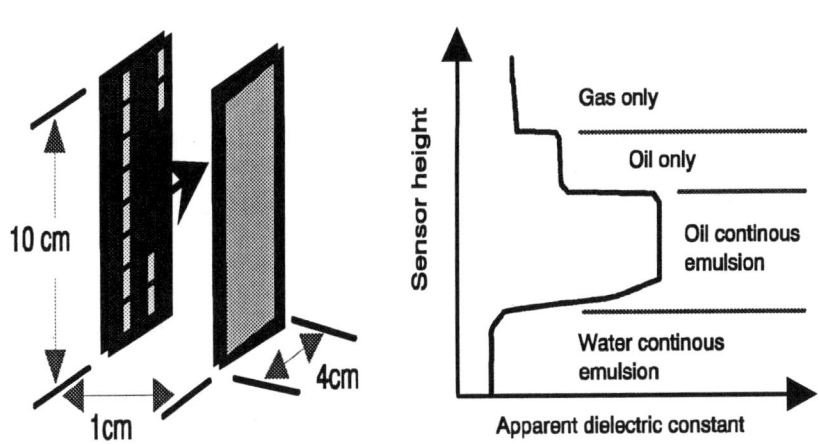

FIGURE 2.

**MEASUREMENT PRINCIPLE - MCF SYSTEM**

The MCF meter, which has been developed by KOS in cooperation with A/S Norske Shell and Shell Research, has been successfully tested in multiphase laboratories as well as in several field locations in Oman and Gabon. Test results show that the meter is capable of measuring multiphase flow with an accuracy of better than +/- 10 % per phase over a wide range of flow conditions. During the laboratory and field tests the meter has been subjected to both sand and wax production. This does not adversely affect the reliability or performance of the meter. Wax removal requirements will be as for other components along the conduit.

The sensor plates, the surrounding electronic hardware and the software package for flow metering and analysis have been proven in onshore laboratories as well as in several field applications. Putting an MCF subsea will not affect the measurement technology of the MCF system. The principals are the same, the flow through the subsea pipes are not different from the flow onshore, exept prehaps for higher pressures, larger pipe diameters and more cooling.

## MULTIPHASE METERING GOING SUBSEA

Putting a multiphase flow meter subsea will require due care and attention if successfull introduction and use are to be achieved.

### MCF System configuration

The MCF system configuration has been defined with the aim of minimizing the changes required for going subsea. Key considerations have been the installation requirements for the sensor arrangement, the design of the electronic circuitry and processing systems, and flexible communication interfaces.

The sensor is small in size and maintains sufficient strength and reliability, while the electronic circuitry, signal handling, data processing, communication interfaces etc. offers low power consumption, high reliability and availability, substantial data storage capacity and a number of data communication possibilities.

These are benefits equally valuable to both the onshore version of the MCF and the Subsea MCF. The onshore version requiring solar powering, modem type interface to remote operator station, low maintenance requirements etc., all features (exept for solar powering) also required by the Subsea MCF. **Refer Figure 3.**

Using a Subsea MCF system will result in major changes compared to conventional technology. Typically, test flowlines and test manifolds may be removed, which in turn will result in substantial cost savings. This is illustrated by the overview of a typical subsea configuration with and without multiphase metering as given in **Figure 4.**

### Subsea Requirements

Taking a multiphase flow meter subsea introduces requirements not normally applied to multiphase flow meters, largely related to the following key issues:

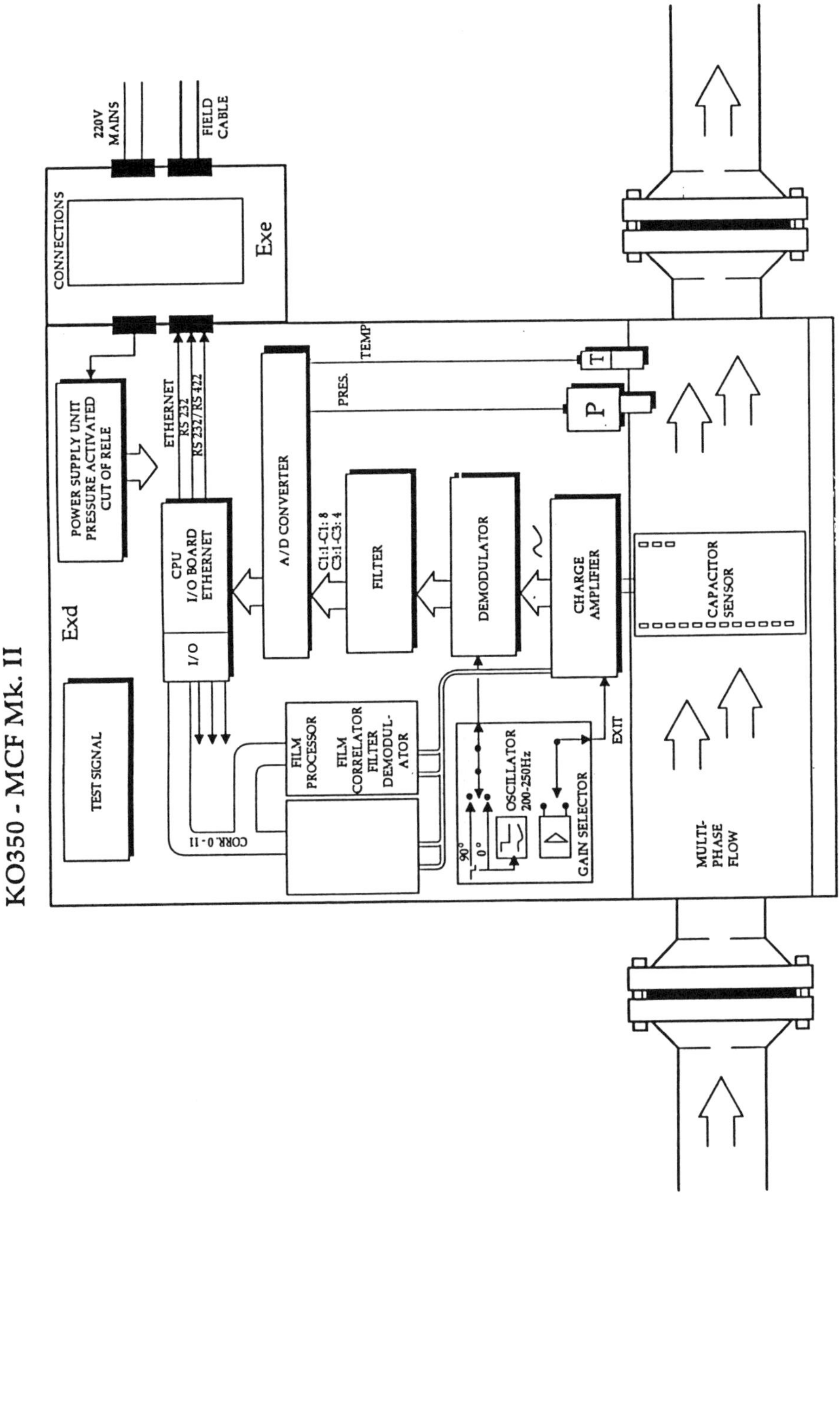

# KO350 - MCF Mk. II

**FIGURE 3.    BLOCK SCHEMATIC DIAGRAM - MCF SYSTEM**

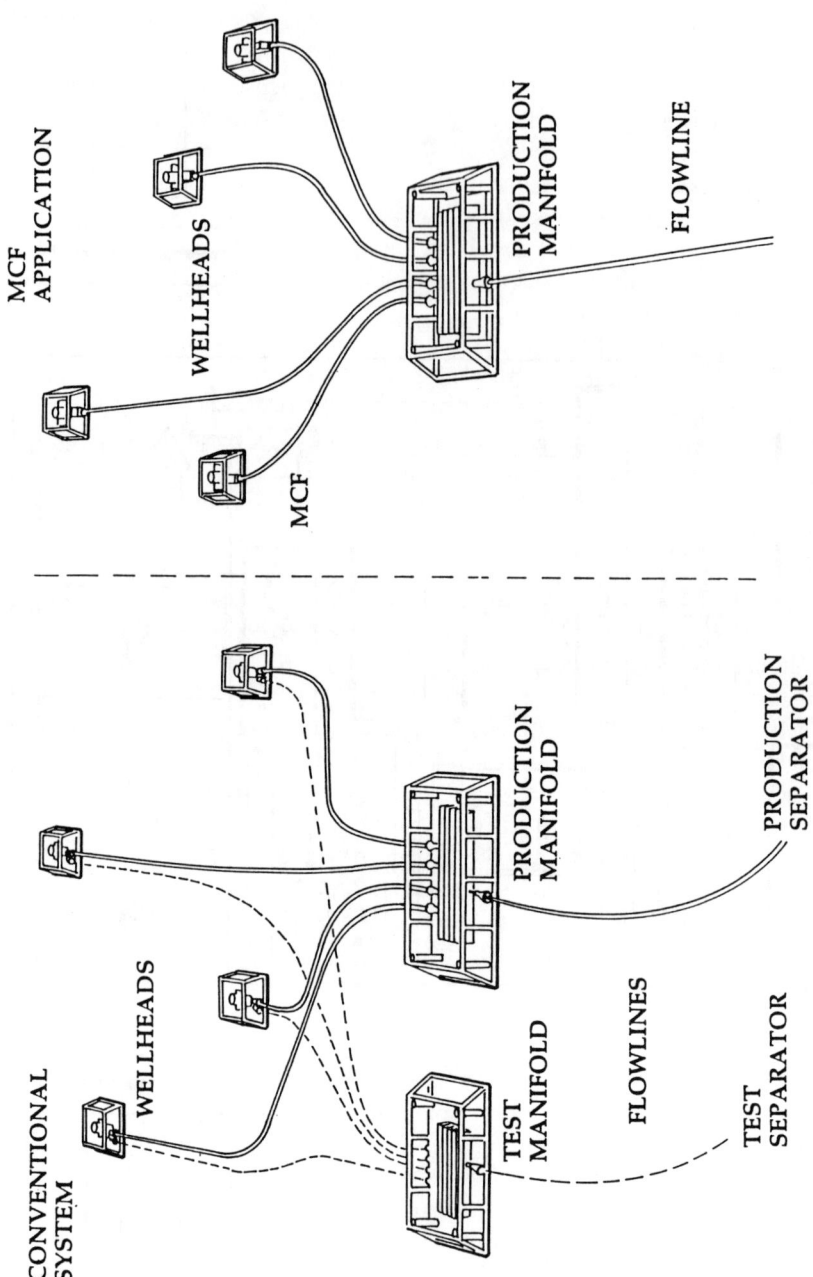

FIGURE 4.
SUBSEA PRODUCTION
CONVENTIONAL APPLICATION versus MCF APPLICATION

- Standards and regulations for subsea equipment (typically API 6A and API 17D related to subsea wellhead and christmas tree equipment)
- Space limitations may prevent flow conditioning. Upstream and downstream flowlines will have limited lengths due to limited size of subsea structures.
- Access is limited and may be difficult. Typically, flowlines are normally running below protective skirting etc. reducing the available space for instalation of meters. Particularly when it comes to non permanent meters.
- Electronic containers etc. must be suitable for applicable water depths. This in turn may increase weight and size due to the thickness required for electronic housings as waterdepths gets greater.
- Considerations must be made with regards to reliability, availability and possible requirements for redundancy. Today most subsea control systems depend on dual redundant electronic configuration with a design life of 25 years.
- Power supplies and communication interfaces should be limited, and if possible combined with existing subsea equipment. The possibility for sharing communication interfaces with existing subsea control systems should be encouraged, as additional communication lines will increase the cost of the umbilicals. Similarly, power consumption should be kept to a absolute minimum. This again to minimize the cost impact on umbilicals. In addition the communication interface must be suitable for interface to subsea control systems from all major manufacturers.
- Installation arrangements must be considered for both permanent and insert type arrangements. If the field has a lifetime expectancy of less than five years it may be acceptable to install meters on a permanent basis. However, most fields will have a much longer lifetime expectancy, and in those cases, retrievable arrangements should be chosen.
- The use and advantages of Remotely Operated Vehicle (ROV) based installation should be carefully compared to installation by Running Tool and Guidepost arrangements. ROV installation will offer substantial advantages when it comes to installation and retrieval operations. However, using an ROV does put stringent limitations on the design of the meters.
- Whenever it is not possible to use an ROV, i.e. access limitations etc., a Running tool for Guide wire installation should be considered. For some installations the space limitations are critical. For this case the use of running tool might be cost effective, especially if divers can perform inspection and cleaning tasks and operate the tool. The disadvantage of Running tool and Guide post installation is additional cost for tool arangements and also installation and retrieval equipment on the surface vessel.
- Material selection, surface treatment and corrosion protection must take into account both the well fluids and the environmental conditions, often while ensuring that weight can be maintained at a minimum. In particular the weight is of importance when it comes to design of ROV based installations. For these type of installations, leight weight materials such as Titanium should be considered.

- Testing of the system in a onshore flow laboratory and as a part of the total subsea system must be carefully planned and evaluated. As for any other flow meter it is important to verify that the chosen design does meet with the necessary flow measurement accuracy requirements. Similarily, it is vitally important to ensure that necessary vertification testing is included for the components going subsea. Testing in shallow waters should be included to verify installation and retrieval as well as Running tool interface etc. where Running tool is being used.

In addition to the design requirements given, a number of operational requirements must also be considered. For some time it has been a key target to remove test flowlines and test separation equipment when installing multiphase flowmeters. This does however introduce other problems. Typically; how do you perform subsea well sampling without a test flowline; and similarily, can it be accepted that the production redundancy offered by the test separator is no longer available.
A number of other production considerations will also apply.


### The KOS Subsea MCF
KOS initiated the development of Subsea version of the MCF multiphase flowmeter in the autumn of 1993, following the successfull field tests of the onshore version of the meter. The development has been aimed at producing an ROV installable multiphase flowmeter, utilizing the measurement technology developed in the onshore versions of the meter.
No Subsea MCF meter has yet been produced, but the design requirements have been established and the manufacturing documentation is prepared now. The first meters are expected to become commercially available by end 1995.

Three different installation methods are considered relevant; ROV carried installation, topside suspension with ROV horizontal coarse guiding, or running by a Running tool on guide wires. **Refer Figures 5a, 5b and 5c.**


### Subsea MCF - Design configuration
The Subsea MCF will be suitable for installation on a christmas tree structure, on a flowline and / or on a subsea manifold. Actual installation location will be dependant on flow conditions (and available space).
The meter will be installed into the subsea system at the construction site.

To achieve good retrievability of critical components for eventual service, the Subsea MCF will be made as a retrievable insert component, comprising all electronics and sensor elements and a permanently installed housing (spool piece) with guidance structure. **Refer Figures 6** for an artists impression on the possible use of a Subsea MCF, and **Figure 7** for a drawing of the current design of the Subsea MCF.

FIGURE 5a.

INSTALLATION METHODES - SUBSEA MCF
ROV INSTALLATION

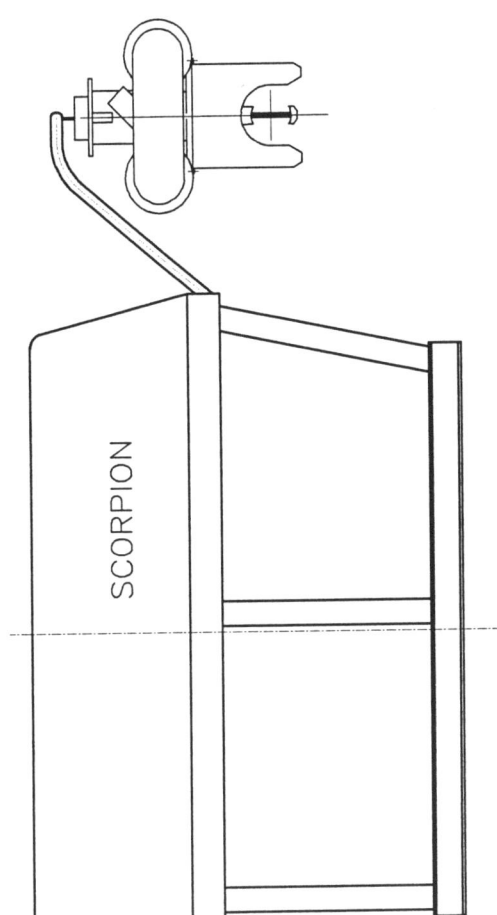

SCORPION

FIGURE 5b.

INSTALLATION METHODES - SUBSEA MCF
TOPSIDE SUSPENSION WITH ROV GUIDANCE

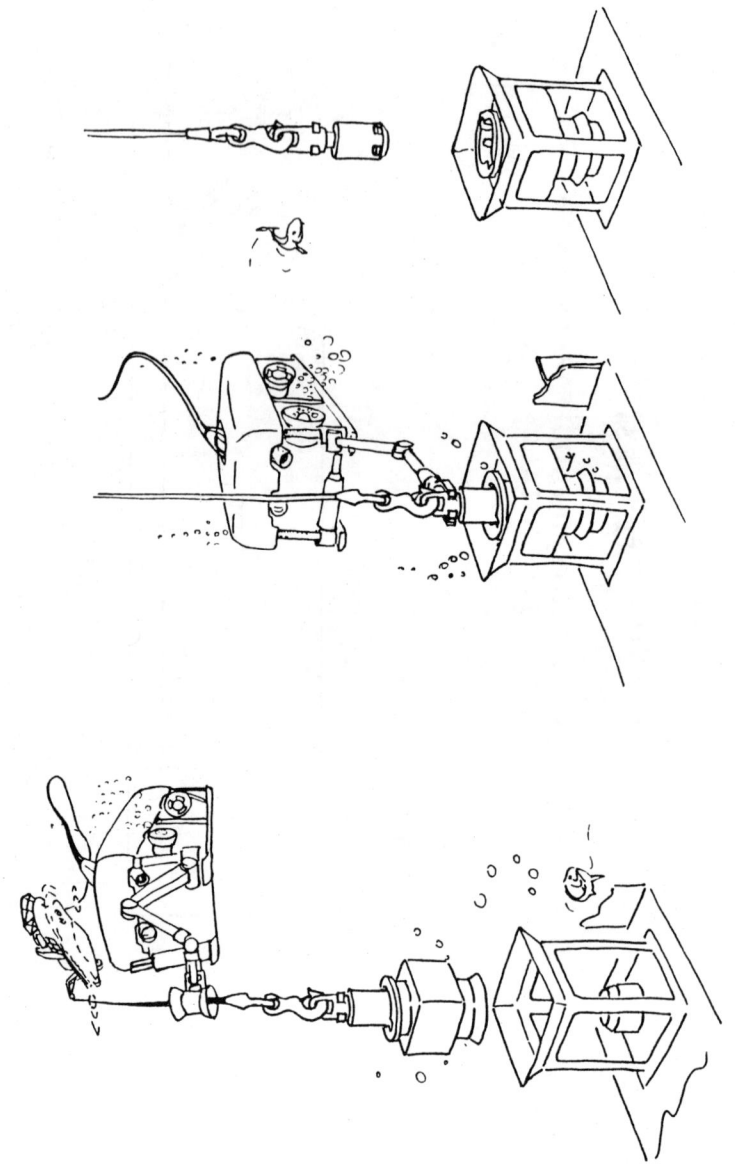

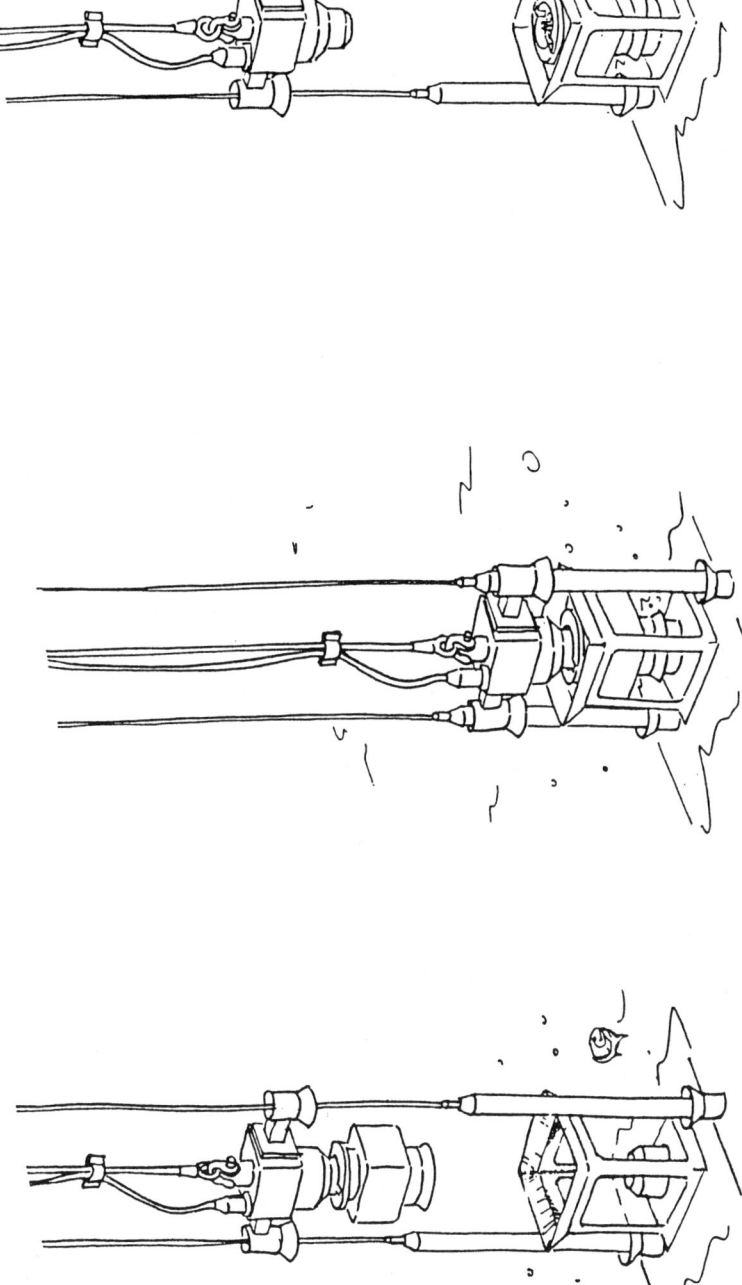

**FIGURE 5c.**

**INSTALLATION METHODES - SUBSEA MCF RUNNING TOOL ON GUIDE WIRES**

In going for a ROV based system, KOS has set the requirements for the maximum meter size and weight. Today most ROV manipulators can handle a maximum of 50 kg in water. This weight limit is also applicable if a stinger arrangement is used on the ROV (extra boyancy can then be avoided). This means that the Subsea MCF meter must have a wet weight of less than 50 kgs. In addition the weight in air should not exceed 100 kg to enable efficient handling. This has also been the design requirement for the Subsea MCF meter.

To achieve this the connector between insert and spool piece must have low weight design. Similarily, it will not be possible to use a large metering device, nor a large electronic housing.

Using an ROV based system simplifies operational limitations, in that the meter can be easily installed and retrieved without having to bring in a sophisticated barge or special intervention vessel with necessary tools etc. When using an ROV, a monohull ship shaped vessel or other type of vessel equipped with an ROV spread, can accomplish the task of retrieval and installation.

In the event a sensor insert has been retrieved for service, it is important to clean and inspect the subsea receptical arrangement prior to reinstallation. For this purpose, an ROV mounted suction pump or jet can be used to clean and remove debris. A special cleaning tool will be made to clean up the precious metal sealing surface of the spool piece subsea. This will ensure that the sealing surface is well prepared to ensure the performance of the reinstated metal to metal seal. The metal to metal seal seal will be dimensioned for 1.2 times sea water pressure at the installation depth (water pressure from the outside), and for hydrostatic test pressure from the bore.

Further, the spool piece will have no moving parts, and the pressure drop across the spool piece will be kept to a minimum. Corrosion protection will be provided by choice of materials (and by use of sacrificial anodes where necessary).

The electronics will be mounted in a 1 atm. nitrogen filled housing to ensure long life time for electronics. To minimize the risk of intrusion of sea water to the electronics, there will be two pressure barrieres between the athmospheric housing and the sea wherever practical.

Low power consumption is important to minimize design impact on the control umbilical. To enable low power consumption while maintaining a self sufficient electronic measurement and processing system subsea, high reliability surface mounted electronics will have to be applied. The number of print circuit boards in the system will be kept to an absolute minimum. The goal is that the power can be supplied through the same cable cords as for the subsea control pods.

The Subsea MCF will have facilities for local data processing, data storage and communication line interface. Accumulated measurement values can be transferred topside on request. This will minimize the signal traffic and thus the band width requirement for the signal cords. The system has been designed with flexible communication possibilities enabling interface to most subsea control systems in parallel with the pods as a stand alone unit.

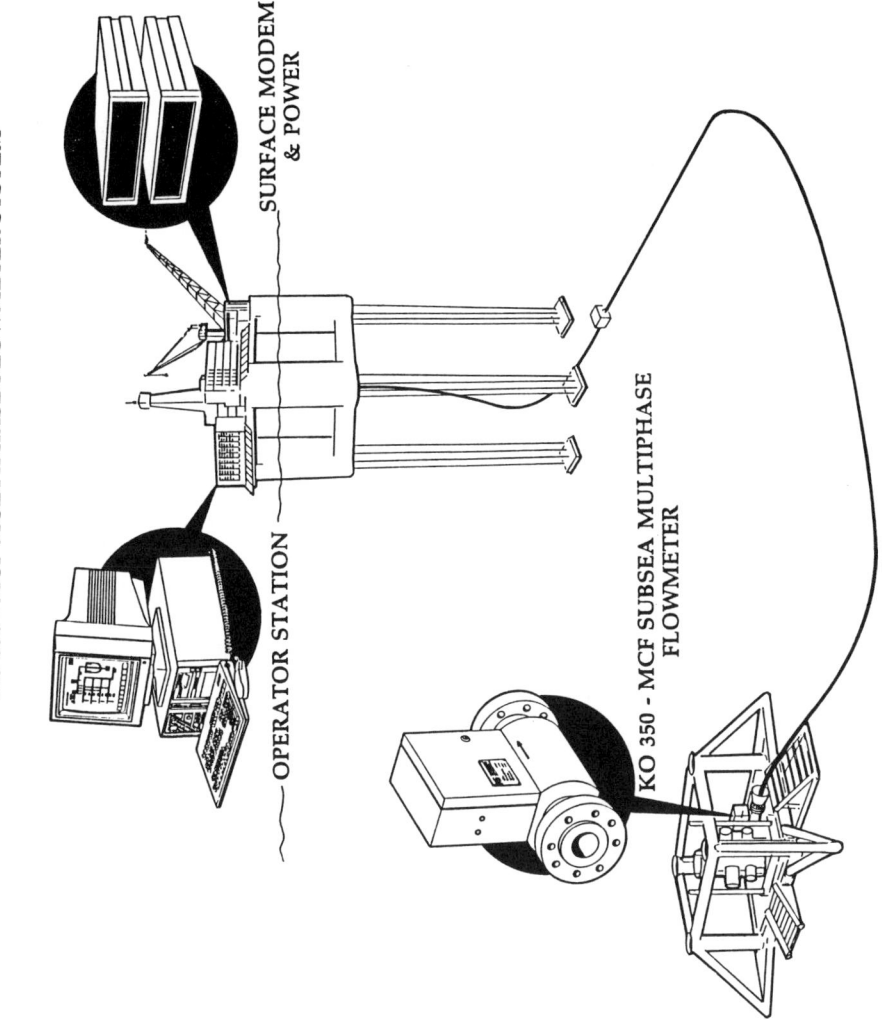

FIGURE 6.

SUBSEA INSTALLATION
KO 350 - MCF MULTIPHASE FLOWMETER SYSTEM

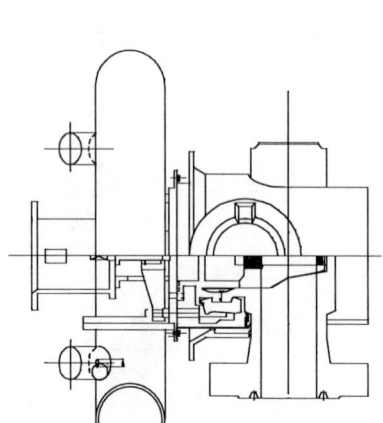

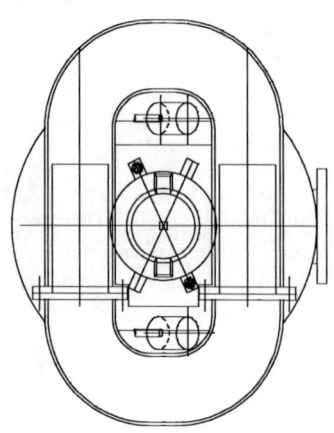

FIGURE 7.
SUBSEA MCF - GENERAL ARRANGEMENT

## The Subsea MCF Sensor arrangement

A critical component for a Subsea MCF will be the insert sensor arrangement. Guiding the sensor plates into position without damage is vitally important. The use of a combined protection/guiding skirt will be applied to protect the sensors during transport, deployment and retrieval. Coarse positioning and coarse rotational alignment of insert above guide funnel will be performed by ROV (N/A for running on guide wires).

The effect of wax built-up on the plates will be monitored and corrected for by the Subsea MCF system and alarmed if necessary. The Subsea MCF will have no facilities for cleaning of wax build up, and will therefore be dependent on a chemical injection system (wax inhibitor) upstream (X-mas tree or other) the meter.

Possible wear or tear of the sensor plates caused by sand erosion or other effects will also be monitored and corrected for by the Subsea MCF system. Any serious wear of the sensor plates will be alarmed, and will give indications with regards to e a necessary replacement of the sensor plates.

As the flow sensor is intrusive, there will be no possibilities for pigging of the flowline after installation. However, pigging can be performed with a dummy insert, without sensor plates in situe.

## Other design considerations

All calibration of the MCF meter will be performed topside / onshore, as subsea calibration facilities will not be available. Adjustment of the initial calibration values will be based on samples taken from each well. This will ensure that the number of interventions and maintenance adjustments are kept to a minimum.

Note however that a subsea sampling system is now being considered by KOS to enable determination of critical calibration values for the Subsea MCF, as well as to give the operator the necessary access to well flow samples for reservoir monitoring etc.

Another very important requirement is to be able to download new equations and algorithms into the subsea processor facilities. This will enable improvements of the system resulting from new developments (typically extended flow range, improved accuracy etc.) without having to retrieve the Subsea MCF.

Connection, disconnection and monitoring operations during intervention shall be performed by ROV. Seal in well flow will be tested externally from ROV stab after connection. External electrical connector(s) will be both ROV and diver operable and the insert will have a contingency release mechanism that can be operated by ROV.

The Insert will be easily adaptable to use of a tool run on guide wires for installation and retrieval at diver accessible water depths. This will ensure that only minor changes will have to be performed for non ROV installations. However, the cost and availability of a running tool must be considered.

It is recommended that a purge facility is included in the permanent installation, but not as part of the MCF system. This to ensure that the flow line past the insert can be cleaned prior to retrieval such that well fluid is not released to the environment.

The external subsea system shall have isolation valves both upstream and downstream of the Subsea MCF. This to enable closure of lines and bleed off of pressure prior to connection / disconnection of insert arrangement.

## DISCUSSION

The development of the Subsea MCF has been based on field proven technology from both onshore testing of the meters and from previous subsea use of key components in the chosen design.

Operational experience being fed back to the KOS organization from both the onshore field installations and from existing subsea installations of similar equipment, will continue to help the development and future improvement of both the onshore and the Subsea MCF systems.

The use of Subsea MCF systems in new subsea developments are being discussed with major oil companies, and the first installation will be regarded as a key milestone for the use of multiphase flow meters subsea.

KOS hope to bring the first Subsea MCF on stream by mid 1995, which should ensure commercial availability by end 1995

## REFERENCES

/1/     D. Brown, J.J.den Boer and G. Washington: "A multi-capacitor multiphase flow meter for slugging flow", The North Sea Flow Measurement Workshop, October 1992.

/2/     E.H. Lunde: "Multiphase metering", Subsea Engineering News,  April 1993.

/3/     C.J.M. Wolff: " The required operating envelope of multiphase flowmeters for oil production measurement", The North Sea Flow Measurement Workshop, October 1993.

/4/     D. Brown: "Field experiece with the multi-capacitor multiphase flow meter", North Sea Flow Measurement Workshop, October 1993.